THE
DESIGN
THINKING
PLAYBOOK

Published by John Wiley & Sons, Inc., Hoboken, New Jersey.
Published simultaneously in Canada.

For general information on our other products and services or for technical support, please contact our Customer Care Department within the United States at (800) 762–2974, outside the United States at (317) 572–3993 or fax (317) 572–4002.

Wiley publishes in a variety of print and electronic formats and by print-on-demand. Some material included with standard print versions of this book may not be included in e-books or in print-on-demand. If this book refers to media such as a CD or DVD that is not included in the version you purchased, you may download this material at http://booksupport.wiley.com. For more information about Wiley products, visit www.wiley.com.

Library of Congress Cataloging-in-Publication Data

Names: Lewrick, Michael, author. | Link, Patrick, 1968- author. | Leifer,
   Larry J., author.
Title: The design thinking playbook : mindful digital transformation of teams, products, services,
   businesses and ecosystems / by Michael Lewrick, Patrick Link, Larry Leifer.
Description: Hoboken : Wiley, [2018] | Includes bibliographical references and index. |
Identifiers: LCCN 2018011184 (print) | ISBN 9781119467472 (pbk.)
Subjects: LCSH: Creative ability in technology. | Creative ability in business. | Creative thinking. |
   Industrial management–Technological innovations. | Technological innovations.
Classification: LCC T49.5 .L49 2018 (print) | DDC 658.4/094–dc23
LC record available at https://lccn.loc.gov/2018011184

ISBN 978-1-119-46747-2 (pbk)
ISBN 978-1-119-46748-9 (ebk)
ISBN 978-1-119-46749-6 (ebk)
ISBN 978-1-119-46750-2 (ebk)

Design and layout: Nadia Langensand
Cover design and visualization: Nadia Langensand

Printed in the United States of America

V10007899 012919

THE DESIGN THINKING PLAYBOOK

MINDFUL DIGITAL TRANSFORMATION
OF TEAMS, PRODUCTS, SERVICES,
BUSINESSES AND ECOSYSTEMS

BY:

MICHAEL   PATRICK   LARRY
LEWRICK   LINK   LEIFER

VISUALIZATION: NADIA LANGENSAND

WILEY

Content

The Tetris blocks will guide you through *The Design Thinking Playbook*. We start with a better understanding of the individual phases of the design thinking cycle. In the thematic block of "Transform," we discuss the best ways to shape framework conditions and how strategic foresight helps us to create greater visions. The last part, "Design the Future," focuses on the design criteria in digitization, the design of ecosystems and the convergence of systems thinking and design thinking, and the options of how to combine data analytics and design thinking.

DESIGN THE FUTURE

1. UNDERSTAND DESIGN THINKING   13

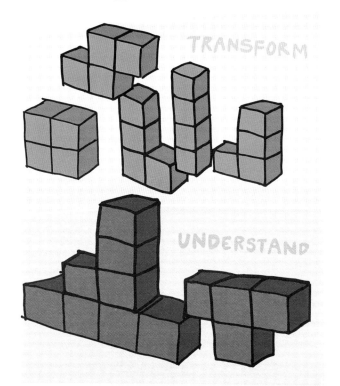

TRANSFORM

UNDERSTAND

3. DESIGN THE FUTURE 211

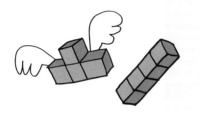

2. TRANSFORM ORGANIZATIONS 131

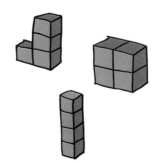

Keep up the design thinking mindset and start hunting for the next big opportunity!

DESIGN THINKING
BEYOND DIGITAL

RADICAL INNOVATION

Foreword

Prof. Larry Leifer

- *Professor of Mechanical Engineering Design, Stanford University*
- *Founding Director, Stanford Center for Design Research*
- *Founding Director, Hasso Plattner Design Thinking Research Program at Stanford*

HELLO

I am quite delighted with this collection of factors of success in design thinking. My special thanks go to Michael Lewrick and Patrick Link. I also want to thank Nadia Langensand, who is responsible for the artistic implementation. As an interdisciplinary team, we were able to put together a fantastic book.

I want to express my gratitude as well to all the experts who shared their knowledge with us and contributed to reflections on the subject matter. The book that emerged is not only one on design thinking but also an exciting essay with deep insights into the application of design thinking beyond the digital context. This *Playbook* is entertaining and motivates readers to do it, not just think about it.

The book stimulates and helps readers

- to put well-known and new tools in the right context of the application;
- to reflect on the entire scope of design thinking;
- to direct the necessary mindfulness to the personas of Peter, Lilly, and Marc;
- to accept the challenges of digitization in which new design criteria in the human–machine relationship, for instance, are increasingly gaining in importance; and
- to set an inspiring framework in order to enshrine design thinking even more strongly in our companies and generate radical innovations.

I'm delighted in particular that this book contains contributions from experts from actual practice as well as from the academic world. A few years ago, the idea took shape that a stronger networking of the design thinking actors should be achieved. This *Playbook* and the communication in the Design Thinking Playbook (DTP) community today act as a stimulus for an open exchange of ideas and materially contribute to enshrining design thinking and new mindsets in companies.

Design thinking is currently experiencing a surge of interest because it is a pivotal tool for initiating digital transformation. We have seen how banks use it to shape the "era of de-banking" and how start-ups have created new markets with business ecosystem design.

At the ME310, which has become somewhat of a legend by now, I frequently have the honor of welcoming corporate partners from different industries from all over the world, who work out exciting design challenges with our local and international student teams.

Have fun reading!

Larry Leifer

DTP - COMMUNITY
WWW.DT-PLAYBOOK.COM

Driven by curiosity

We are curious, open, ask "WH questions" continuously, and change the perspective in order to look at things from various sides.

Focus on the people

We focus on the human being, build empathy, and are mindful when exploring his/her needs.

Accept complexity

We explore the key to complex systems, and accept uncertainty and the fact that complex system problems demand complex solutions.

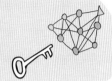

Visualize and show

We use stories, visualizations, and simple language to share our findings with the team or create a clear value proposition for our users.

Experiment and iterate

We build and test prototypes iteratively to understand, learn, and solve problems in the context of the user.

THE DESIGN THINKING PLAYBOOK
MINDSET

Co-create, grow, and scale

We continuously expand our capabilities to create scalable market opportunities in a digital world and in ecosystems.

With varying mental states

As the situation requires, we combine different approaches with design thinking, data analytics, systems thinking, and lean start-up.

NEW MINDSET.
NEW PARADIGM.
BETTER SOLUTIONS.

WWW. DESIGN-THINKING-PLAYBOOK.COM

Develop process awareness

We know where we stand in the design thinking process and develop a feeling for the "groan zone" to change the mindset through facilitation in a targeted way.

Networked collaboration

We collaborate on an ad hoc, agile, and networked basis with T-shaped people and U-shaped teams across departments and companies.

Reflect on actions

We reflect on our way of thinking, our actions, and attitudes because they have an impact on what we do and on the assumptions we make.

Introduction

Where will the next major market opportunities emerge?

Here   Here   Here   Here   Here   Here

Done!

The hunt for the next big market opportunity is ever present. Most of us are ambitious company founders and good employees, managers, product designers, lecturers, or even professors. We all had cool business ideas at one time, such as the vision to build a revolutionary social network 3.0 that would outshine Facebook, to establish a health care system that suggests the best possible treatment option to patients and gives us health data sovereignty, and many more ideas.

It's people like us who develop new ideas with great energy and tireless commitment, our heads filled with visions. To succeed, we usually need a (customer) need; an interdisciplinary team; the right mindset; and the necessary leeway for experimentation, creativity, and the courage to question what exists.

Across all sectors, it has become increasingly important to identify future market opportunities and enable the people inside organizations to work with agility and live creatively. Today's planning and management paradigms are frequently not sufficient to respond appropriately to changes in the environment. Moreover, many companies have banned creativity in favor of operational excellence and management-by-objectives.

The old management paradigms must therefore be dissolved. But they will dissolve only when we allow new forms of collaboration, apply different mindsets, and create more room to develop and find solutions.

What are the three things that are important to us?

1) Keep your personality!

"No one needs to mutate into Karl Lagerfeld just because we have creativity and room for development at our disposal!"

Because we are people with different personalities, it is vital we remain who we are today and continue to trust our experience and intentions in order to implement what we have gotten off the ground so far. And if there is one thing we have learned from Tetris, it is the fact we too often try to fit in somewhere—with the unfortunate consequence that we then disappear!

2) Love it, change it, or leave it!

*"Use the concepts and tips that you like
and adapt them to your needs."*

We decide on our own which mindset fits our organization, and whether we like the expert tips in this book or find them absurd and want to change them or adapt them to our situation. It would be a shame if all organizations were a clone of Google, Spotify, or Uber. Each company has its own identity and values. Even in Tetris, we have the possibility of turning things around at the last second so we're successful in the end.

3) Don't do it alone!

*"Get the necessary skills, technologies, and attitudes on your teams
to be successful and think in business ecosystems."*

We cannot develop products today with the mindset, design criteria, and needs from the past. Both users' needs and the way we work together have changed, and we must have the necessary freedom and the skills to develop products and services, business models, and business ecosystems with agility in a digitized world. If we do not transform our organization, failed growth initiatives will pile up.

LOVE IT! CHANGE IT! LEAVE IT!

TIME FOR CHANGE!

What should you expect?

The Design Thinking Playbook will help you in the design of your transition to a new management paradigm. We all know such transitions from our customers' needs. Let's take as an example the transition from analog telephony to the smartphone all the way to a mindphone. While in the 1980s we only occasionally needed to take work calls at home, today we need to be reachable everywhere and anytime. In the future, we might want to control simple communications directly through our thoughts in order to eliminate the inefficient manual-entry smartphone interface. Successful companies have also created business ecosystems in which they closely integrate customers, suppliers, developers, and hardware manufacturers.

In this *Playbook*, we make the world of design thinking palpable—and we want to see you a little happier at the end! Because design thinking also creates happiness. And when you as readers are happy, we have been successful!

What was the biggest challenge?

"Do you actually know the needs of your readers
for whom you're writing the book?"

(Source: Direct quote from the first meeting of the editors and contributors of the *Playbook*)

Although we could very well picture ourselves as potential readers of such a playbook, we complied with the wish expressed in the question. In the design thinking style, we first determined the customer needs, created various personas, and developed much empathy for the work of our colleagues in order to establish a solid basis. Thus *The Design Thinking Playbook* is the first design thinking book that lives the mindset from the first to last page!

Because there is already a lot of design thinking literature on the market, we felt the need to show how design thinking is used optimally. We also want to help you professionalize your design thinking skills. And because the world does not stand still, we reflected on the digital paradigm and combined design thinking with other mindsets with the aim of becoming better and more innovative in a digitized world.
Enough of our introduction. Let's focus on what's essential—the specific and practical application of design thinking and expert tips. We tried to formulate the tips as comprehensible activities and ways of working. The "How might we . . ." instructions provided are supposed to be no more than indications for how we can proceed. Design thinking is not a structured process! We adapt the mindset and the approach to the particular situation.

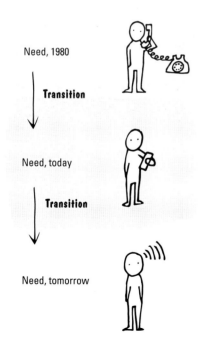

Need, 1980

Transition

Need, today

Transition

Need, tomorrow

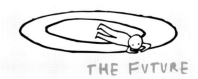

THE FVTVRE

1. UNDERSTAND DESIGN THINKING

"Who is Peter?"

As described in the Introduction, we wanted to write a book for all those interested in innovation, for movers and shakers as well as entrepreneurs who design digital and physical products, services, business models, and business ecosystems as part of their work. Regarding our three personas, we were able to identify three very different kinds of users who apply design thinking in their day-to-day activities. One thing the three have in common, though: All three of them want to create something new in a rapidly changing world. Which brings us straight to our initial question:

How can we learn more about a potential user and better uncover his or her hidden needs?

In the individual chapters, we focus on the three personas of "Peter," "Lilly," and "Marc." We hope this lets us address the needs of design thinking practitioners as best as possible.

Peter, 40 years old, works at a large Swiss information and communications technology (ICT) company. Peter came in contact with design thinking in the context of a company project four years ago. Peter was a product manager then. Searching for the next major market opportunity, he had already tried out quite a few things. For a while, Peter always wore red underwear on New Year's Day, but it didn't make him any luckier in terms of successful innovations. After this experience, he doubted at first whether design thinking was really something for him. It was hard for him to imagine that something useful could come out at the end of the described procedure. The approach seemed just a little esoteric to him.

His attitude changed after he attended a number of co-creation and design thinking workshops with customers. He felt the momentum that can come into being when people with different backgrounds tackle complex problems together in the right environment. Paired with a good facilitator who provides work instructions in a targeted manner, any group can be empowered to create a new experience for a potential user. This positive experience prompted Peter to take on the role in such design thinking workshops as a facilitator.

Owing to the workshop experience he had gotten and its successful implementation in projects, Peter was promoted not long ago. He now has the privilege of calling himself an "Innovation & Co-Creation Manager."

He is glad to meet like-minded spirits at events such as "Bits & Pretzels" in Munich or design thinking meetings in Nice, Prague, or Berlin where he can exchange thoughts and ideas with the who's who among digitization evangelists.

More about Peter: What is his background experience?

Peter studied at the Technical University of Munich. After graduating, he held various positions in the telecommunications, IT, media, and entertainment industries. Five years ago, he decided to move from Munich to Switzerland. Its location and excellent infrastructure convinced him to make this daring change. There, Peter met his future wife, Priya. He has been happily married for two years. She works for Google at their corporate campus in Zurich. Priya is not allowed to talk much about the exciting topics she works on, although Peter would be quite intrigued by them.

Both like to get involved with new technologies. Be it the smart watch, augmented reality, or using what the sharing economy has to offer, they try out everything the digital world comes out with. A few weeks ago, Peter had his dream of getting a Tesla come true. Now he is waiting for his car to be self-driving soon so he can enjoy the beautiful landscape while looking out the window. In his new role as Innovation & Co-Creation Manager, Peter now belongs among the "creative" ones. He has replaced his suits and leather shoes with Chucks.

Peter tried to resolve the last crisis in his relationship with a little design thinking session. Priya was very aloof with Peter all of a sudden. Peter took the time to listen to Priya and better understand her needs. Together, they discussed ways to bring more oomph into their relationship. During the brainstorming, Peter had the idea that wearing his lucky red underwear might save the relationship. But in the meantime, he had developed so much empathy for Priya's concerns that he quickly dismissed the idea. In the end, they came up with a couple of good ideas for their relationship. Priya did wish, though, that Peter would use a different method to learn his needs besides design thinking.

Up to now, Peter had used design thinking in various situations. He learned that the approach sometimes worked very well for reaching a goal, but that sometimes it wasn't right. He would like to get a couple of tips from experienced design thinkers to shape his work even more effectively.

Visualization of the persona

Leading a team

Design workshop facilitator

Peter, Innovation & Co-Creation Manager

Would like to be a design thinking expert

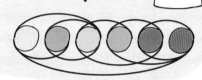

Creative

Analytical

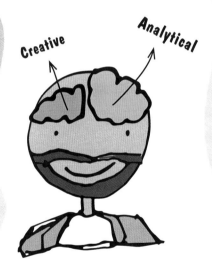

Development of new product, process, and service ideas in the information and communications industry

Building up community & exchange of knowledge with other design thinkers

MSc in Electrical Engineering, Technical University, Munich. ICT, media and entertainment

Developing digital business models and implementing digitization strategies

Pains:

- Peter's employer does not invest much in the further training of employees.
- Although Peter feels quite competent by now in dealing with design thinking, he is still convinced he could get more out of the approach.
- Peter has noted that, while design thinking is a powerful tool, it is not always used optimally.
- Peter frequently wonders how the digital transformation might be accelerated and what design criteria will be needed in the future to be a success on the market.
- Peter would like to combine other methods and tools with design thinking.
- Peter is faced with the challenge of having to impart to his team a new mindset.
- He would like to exchange ideas with other design thinking experts outside his company.

Gains:

- Peter has a lot of leeway in his daily work to try out new methods and tools.
- He loves books and all tangible things. He likes to use visualizations and simple prototypes for explaining things.
- What he would really like to do is establish design thinking in the whole company.
- He knows various management approaches he would like to link with design thinking.

Jobs-to-be-done:

- Peter has internalized the design thinking mindset., but sometimes, good examples that would help to change his environment don't come easily to him.
- Peter enjoys trying out new things. With his engineering background, he is open to other approaches to problem solving (whether quantitative or analytical).
- He would like to become an expert in this environment as well. He is looking to connect with like-minded individuals.
- Peter experiments with design thinking.

Use cases:

A book in which experts report on their experience, in which tools are explained by way of examples—such a book would be just the thing in Peter's eyes. A book he could recommend to his company at all hierarchical levels. A book that expands the framework of inspiration and makes people want to learn more about design thinking. He would also like to know which design criteria will be needed in the future, in particular for the development of digital products and services.

"Who is Lilly?"

Lilly, 28 years old, is currently working as a design thinking and start-up coach at Singapore University of Technology & Design (SUTD). The institute is one of the pioneers in design thinking and entrepreneurship for technology-oriented companies in the Asian region. Lilly organizes workshops and courses that combine design thinking and lean start-up. She teaches Design Thinking and coaches student teams in their projects. In tandem with that, she is working on her doctoral thesis—in cooperation with the Massachusetts Institute of Technology—in the area of System Design Management on the subject of "Design of Powerful Business Ecosystems in a Digitized World."
To divide participants into groups, Lilly uses the HBDI® (Herrmann Brain Dominance Instrument) model in her design thinking courses. Productive groups of four to five are formed this way, each one of which works on one problem statement. She has discovered that it is vitally important in each group to unite all modes of thinking described in the brain model. Lilly's own preferred style of thinking is clearly located in her right half of the brain. She is experimental, creative, and likes to surround herself with other people.
Lilly studied Enterprise Management at the Zhejiang University School of Management. For her master's degree, she spent a year at the École des Ponts ParisTech. As part of the ME310 program and in collaboration with Stanford University, she worked on a project there with THALES as an industrial partner, which is how she became familiar with design thinking. During this project, she visited Stanford three times. She liked the ME310 project so much that she decided to attend the University of Technology & Design in Singapore. There, Lilly became known among faculty for her extravagant flip-flops. SUTD students were less enthusiastic about them.

More about Lilly: What is her background experience?

Lilly has great in-depth theoretical knowledge of various methods and approaches and is able to apply them practically with her teams of students. She is good at coaching these teams but lacks understanding of actual practice. Lilly offers design thinking workshops at the Center of Entrepreneurship at Singapore University of Technology & Design. Frequently, people from industrial enterprises who want to learn more in terms of their innovation capability or better understand the topic of "intrapreneurship" take part in these workshops.
Lilly lives in Singapore and shares an apartment with her friend Jonny, whom she met during her year in France. Jonny is an expat who works for a major French bank in Singapore. At first, Jonny thought Lilly's flip-flops were somewhat freaky but, at this point, he likes that little splash of color on her.
To maximize success, Jonny sees great potential in user-centered design and his bank's pronounced orientation toward customer interaction points. He is enormously interested in new technologies. The thought of what they might mean for banks fascinates and unsettles him at the same time. He follows events in the fintech sector very closely and has identified new potentials that might result from a systematic application of blockchain. He wonders whether such disruptive new technologies will change banks and their business models even more fundamentally than Uber changed the taxi sector or Airbnb the hotel industry—and, if so, when such changes will take place. The core question for Jonny is whether a time will come when banks as we know them will cease to exist altogether. Either way, banks need to become more customer-oriented and make better use of the opportunities that digitization offers than potential newcomers. Jonny is not afraid of losing his job as yet. But still, a start-up together with Lilly might be an exciting alternative. Jonny would like to see his bank apply design thinking and internalize a new mindset, but this is nothing but wishful thinking thus far.

Lilly and Jonny would also like to set up a consulting firm that applies design thinking to support enterprises with digital transformation. They are still looking for something unique that their start-up could offer in comparison to conventional consultancy firms. In particular, they would like to address cultural needs in their approach to consulting. Lilly has observed too many times how the European and American design thinking mindset failed in an Asian context. She wants to integrate local particularities in her design thinking approach: the attitude of an anthropologist, the acceptance of copying competitors, and the penchant for marketing services more quickly, instead of observing the market for a long time. Something else makes them hesitant to implement their plan: They are bit risk averse because next year, once Lilly has completed her doctoral thesis, they want to get married and raise a family. Lilly wants three children.

In her free time, Lilly is active and creative. She often meets with like-minded people she knows from the SkillsFuture program, which is a national program that provides Singaporeans the possibility to develop their fullest potential throughout life, regardless of their starting points; or from events such as "Innovation by Design," which was funded by the DesignSingapore Council. They develop concepts for, among other things, adapting the space and the environment of the country to the needs of people. Lilly is especially intrigued by digital initiatives and hackathons that come into being through real-time data from sensors, social media, and anonymized motion profiles of mobile devices. Singapore is a pioneer that brings the design thinking mindset actively to the entire nation, not least with the "Infusing Design as a National Skill Set for Everyone" campaign.

DESIGN THINKING IN SINGAPORE

Visualization
of the persona

User profile of an experienced design thinker from the academic environment:

**Lilly,
a design thinking &
lean start-up coach**

Research in
the field of agile methods

Set up and maintain
contact
with other design
thinking experts

⇩

Further development of
methods and mindset

Doctoral thesis
"Design of Powerful
Business Ecosystems in a
Digitized World"

Baby
or start-up?

Creative

Analytical

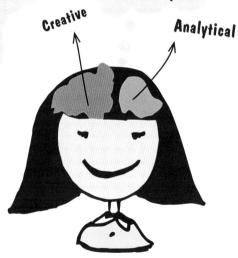

Enterprise Management,
Zhejiang University
School of Management

Is a design thinking
expert

Coaching student
teams

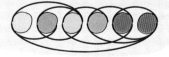

More examples from
actual practice

Ideas Innovation

Pains:

- Lilly is uncertain whether she wants to begin a family or a start-up after she has finished her dissertation.
- Lilly would like to work as a professor in the area of design thinking and lean start-up in Southeast Asia, preferably in Singapore, but no such position exists there yet.
- She feels confident in design thinking both in theory and in her work with students, but she has a hard time establishing its importance for actual practice and convincing partners in the industry of its power.
- Working with colleagues from other departments is difficult, although design thinking could be combined well with other approaches.
- Lilly would like to exchange ideas with other design thinkers throughout the world in order to enlarge her network and make contact with industry partners, but has not yet found a platform to do so.

Gains:

- Lilly enjoys the possibilities offered by the intense contact with students she has as a coach. She can easily try out new ideas, and observing of her students has yielded many findings for her doctoral thesis.
- Lilly loves TED Talks and MOOCs (massive open online classes). She has already attended many courses and talks revolving around the topics of design thinking, creativity, and lean start-up, and has thus acquired a broad knowledge base. She would like to integrate new findings and methods in her courses.
- Lilly wants to bring her knowledge to a community and cultivate contact with other experts, to advance methods, publish, and do research together.
- Through the exchange with those involved in actual practice, Lilly can test and improve new ideas.

Jobs-to-be-done:

- Lilly understands design thinking in theory and is good at explaining the approach to students. But sometimes, she can't think of good new examples and success stories from industry that could motivate the students and workshop participants to try out design thinking on their own.
- Lilly coaches students and start-ups, and organizes design thinking and lean start-up workshops. Her aim is to boost user centricity with all participants.
- Lilly enjoys trying out new things. She knows ethnographic methods and human-centered approaches from her studies. What has surprised her time and again is that the stereotypes of individual disciplines have an element of truth in them, yet interdisciplinary teams still achieve more exciting results.
- Lilly wants to meet new people and find ideas for her work and her start-up.

Use cases:

The book Lilly wants is one that contains many examples and activities from actual practice instead of pure theory. An easy-to-use reference book with tips from experts that widens her inspiration framework and fires her desire for design thinking. A playbook that looks into the future and shows how design thinking will continue to develop. A book that she can recommend to her students as further reading material.

"Who is Marc?"

Marc, 27 years old, completed his MSc in Computer Science two years ago. He used his time at Stanford University to build out his network. He also attended a number of pop-up sessions at the d.school (Stanford University design school) on the theme of entrepreneurship and digital innovation. Marc met like-minded people there who voiced ideas just as crazy as his. Because Marc is somewhat introverted and does not just walk up and speak to people easily, he was grateful for the workshops at the d.school, which were accompanied by a facilitator. The facilitator created an atmosphere in which not only were ideas exchanged but one's thought preferences were recognized, and teams were optimally put together. His group quickly recognized and appreciated Marc as "the innovator." The other team members had knowledge of marketing and sales, finance and management control, and health care and mechanical engineering. The group was thrilled by Marc's idea of stirring up the health care and medtech industry through the use of distributed ledger technology. Marc made quite an impression with words like bitcoin, Zcash, Ethereum, Ripple, Hyperledger Fabric, Corda, and Sawtooth. He waxed enthusiastic about frameworks such as ERIS being miracle weapons to tame the smart-contract dragons. Moreover, Marc had already been involved in two start-ups. For the makers and shakers of two Web analytics firms, he had written code during a summer internship. The group quickly realized that they wanted to found a start-up, knowing very well that Marc's technological affinity for blockchain, together with their business idea, would not yet make a profitable business. Processes and in particular business ecosystems must be designed to initiate a revolution.

More about Marc: What is his background experience?

Marc grew up with mobile communication. As a digital native, he pursues a technology-based lifestyle, as we have already learned. On the level of popular sociology, he is a typical representative of generation Y (why). It is important to him that he do something meaningful with his skills. He wants to work on a team and get recognition. It would be best that no one tell him what to do when it comes to his special field of blockchain.

Marc grew up in Detroit. His parents were middle class. Both of them had made careers in major automotive companies. Hence Marc witnessed how an entire industry can lose its luster bit by bit. The subprime and financial crisis showed him that, from one day to the next, it can become impossible to pay the mortgage for the big mansion in the Detroit suburbs. Marc learned early on how to deal with uncertainties. He internalized how to "dance" with uncertainty and how to weigh options. For him, design thinking and the associated mindset are a natural attitude. Questioning the things that exist and finding new solutions for problems have always been something obvious to him. He owed the privilege of studying at Stanford to a scholarship. Besides the option of founding a start-up with his team from the d.school, he has received job offers at the campus job fair from Spotify and Facebook involving artificial intelligence. Marc likes both companies because they promise he'd be master over his own time and be able to work autonomously.

In his leisure time, Marc is a big baseball fan. His favorite team is the Detroit Tigers.

DETROIT HUSTLES HARDER

Coming home from the job fair in question, Marc met Linda, a Brazilian beauty who works as a nurse at the health center of his university. Marc was so taken by reading on his smartphone about the concept of Everledger that he had stepped onto the bicycle lane. Linda was just able to brake in time, but they both got quite a fright from the encounter. Marc was a bit embarrassed, but then he dared ask Linda if she wanted to network with him on Facebook. Marc is very proud of this. Now they swap emoticons via Whatsapp nearly every hour. Marc usually sends little diamonds—not as a digital assets but as virtual tokens of affection to his lovely Linda. But he was also fascinated by the fact that diamonds as digital assets change owners through a private blockchain.

Visualization of the team

User profiles of a typical start-up team:

Marc, innovator, entrepreneur, tech founder

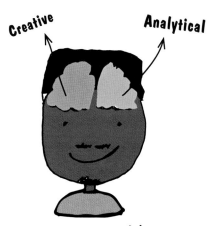

Creative   Analytical

MSc in Computer Science,
Bachelor in Mechanical Engineering
PhD candidate at Stanford University,
Design Research & Innovation

The "brain" (Beatrice)

- creative problem solver
- natural sense for business
- broad knowledge

The mover and shaker (Vadim)

- technical expertise
- implements
- reliable

The visionary (Tamara)

- creative and visionary
- thinks in terms of future opportunities
- sees the big picture

The strategist (Stephan)

- creative and strategic
- thinks in real options
- identifies risks

The salesman (Alex)

- powerful personality
- convincing
- customer-oriented and extroverted

Pains:

- For Marc, his team doesn't learn quickly enough. He wants to conduct simple experiments and develop prototypes in the service environment more rapidly.
- Marc pursues a lean approach for his start-up and has noticed how important it is to be honest with oneself and that the biggest risks should be tested first.
- The dynamics of the market and the technology are so great that even things that have already been tested ought to be questioned again and again.
- Marc always sees new options in the business ecosystem. It is sometimes hard for him to design a complex ecosystem and to shape the business models for the actors in the system.

Gains:

- Marc is enthusiastic about his subject and his team. He enjoys the energizing atmosphere and meaningful work.
- Marc uses design thinking for innovation exchange and combines it with new elements.
- Marc loves the possibilities of digital business models and knows that the whole world is in upheaval, offering start-ups huge opportunities.
- At this point, Marc has come to love interviews and tests with real users. He has learned to ask the right questions and looks forward to the new findings that are spewed forth at a rapid pace.

Jobs-to-be-done:

- Marc wants a book that gives leeway to his natural talent for questioning what exists, presents him with new tools, and shows him how they're applied.
- He wants to know how he can transform his knowledge of information technologies into meaningful solutions. It's vital to him that he find a scalable solution for his blockchain idea quickly and that an innovative business model makes the enterprise viable in the medium term.
- He wishes to work in an environment in which the concept of "teams of teams" is a lived reality, and would like to get suggestions for it.
- With the aid of design thinking, Marc wants to establish a common language and mindset. The dynamics, complexity, and uncertainty are rising. Marc can deal with the situation pretty well, but he has noted that his team is not so good at it.
- Particularly in the blockchain environment, technological development is proceeding quite rapidly. The team must learn from experiments speedily and develop both the market and customers.

Use cases:

Marc would like a book that helps his team adopt the design thinking mindset more speedily and learn faster. The book should contain suggestions and tips, both for experienced design thinkers and for people who are dealing with the mindset for the first time. In addition, Marc would like to get suggestions on how to develop his digital business ecosystem and how to maintain strategic agility even in the growth phase.

EXPERT TIP
Create a persona

How do we proceed when creating a persona?

There are different ways of creating personas. It is important to imagine the typical user as a "real person." People have experience, a life career, preferences, and private and professional interests. The primary goal is to find out what their true needs are. Frequently, potential users are sketched out in an initial iteration, which is based on the knowledge of the participants. It must then be verified that a user who has been sketched out like this actually exists in the real world. Interviews and observations often show that potential users have different needs and preferences than those originally assumed. Without exploring these deeper insights, we never would have found out that Peter likes red underpants and Lilly has a tic with flip-flops. In many workshops, so-called canvas models are used in the context of strategy work and the generation of business models and business ecosystems associated with it. We developed a "user profile canvas" for our workshops that helps in having the key questions at hand and, based upon them, in creating a persona expeditiously.

To promote the creativity of participants and encourage out-of-the-box thinking, it is useful to cut the canvas apart and glue it onto a huge poster. On this poster, the persona can be drawn in full size. In so doing, it is worthwhile to improve the persona iteratively, refining it and digging deeper step by step.

It always makes sense to ask for the "why" in order to get to the actual problem. We try to find out about real situations and real events so as to find stories and document them. Photos, images, quotes, stories, etc., help to make the persona come alive.

In general, work with the persona concept is reminiscent of the procedure applied by so-called profilers (case analysts) in American detective TV series. Profilers are on the hunt for the perpetrators. They solve murders and reconstruct the course of events. They work by describing relevant personality and character traits in order to draw conclusions from behavior.

We recommend taking the time to create a persona yourself. The intensity and closeness are important for building up empathy with the potential user. If time is short, standard personas can be used.

You must be cautious when it comes to personas with brief descriptions. The example of the "persona twins" shows why. Although the core elements are the same, the potential users couldn't be more different. This is why it really makes sense to dig one level deeper to understand the needs in greater detail. We get greater insights, and that makes things even more intriguing.

Persona twins

Prince Charles

Born in 1948
Grew up in England
Married twice
Has children
Successful, rich
Takes vacations in the Alps
Likes dogs

Ozzy Osbourne

Born in 1948
Grew up in England
Married twice
Has children
Successful, rich
Takes vacations in the Alps
Likes dogs

USER PROFILE CANVAS

Name

Persona description

Age, gender, place of residence, marital status, hobbies, leisure time, education and training, position in the company, social environment, Sinus-Milieus category, way of thinking, etc.?

Stories

Stories

Jobs-to-be-done

What task performance is supported by the product?
What are the goals?
Why does it make sense?

Photos

Gains

To what extent do the current products make the customer happy?

Images

Images

Life-size

Use cases

How is the product used, where is it used, and by whom is it used? What happens before and after use?
How does the customer obtain information? What does the purchase process look like?
Who influences the decision?

Photos

Photos

Stories

Pains

What causes a bad feeling in the customer with the current products?
What are the worries of the user?

27

How do we build up empathy with a potential user?

The initial draft of a persona is quickly done. Although just an outline exists, it can be quite helpful and eye-opening. A brainstorming on the team can yield initial insights and contribute to a better understanding; it is absolutely necessary, though, that it be underpinned with real people, observations, and interviews.

In a first step, the user must be defined and found. Ideally, we'll go outside right at the beginning and meet a potential user. We observe him, listen to him, and build up empathy. The insights are well documented, in the best case using photos and videos. If you take pictures, it is important to ask permission beforehand, because not everybody likes to be photographed or filmed! A so-called empathy card can be used here that addresses the following areas: hearing, thinking and feeling, seeing, speaking and doing, frustration, and desire.

We also suggest speaking to experts who know the persona well and, of course, being active yourself and doing what the user is doing.

The credo is: "Walk in the shoes of a potential user!"

Especially when we think we know the products or the situation, we attempt to approach a situation like a beginner—curious and without previous knowledge. Consciously and with all our senses, we go through the experience the user is going through!

After this "adventure," it is useful to define hypotheses on the team, then test them with a potential user or by using existing data, then confirm, discard, or adapt them. The picture of the persona becomes clearer and more solid with each iteration.

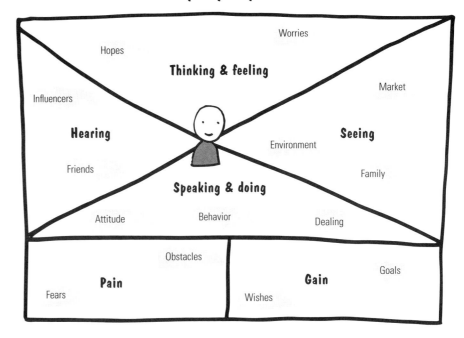

Empathy map

To obtain initial knowledge on the user, another tool that helps is the AEIOU method. AEIOU helps us to capture all the events in our environment.

The task is clear. Get out of the design thinking rooms and speak to potential users, walk in their shoes, do what they do.

The AEIOU questions help to put some structure into the observations. Especially with inexperienced groups, it is easier this way to ensure an efficient briefing on the task at hand.

Depending on the situation, it is useful to adapt the questions individually to the respective observations. The AEIOU catalog of questions and the associated instructions help participants establish contact with initial potential users. Experience has taught us that it helps the groups if a design thinking facilitator or somebody with needfinding experience accompanies first contact of potential users. We all are pretty inhibited when it comes to addressing strangers, observing them, and asking them about their needs. Once the first hurdle has been cleared, some participants and groups develop into true needfinding experts. Chapter 1.4 will deal in greater detail with needfinding and the creation of question maps.

AEIOU is broken down into five categories.
Consider how each of the users behaves in the real world and the digital world.

Activities	What happens? What are the people doing? What is their task? What activities do they carry out? What happens before and after?
Environment	What does the environment look like? What is the nature and function of the space?
Interaction	How do the systems interact with one another? Are there any interfaces? How do the users interact among one another? What constitutes the operation?
Objects	What objects and devices are used? Who uses the objects and in which environment?
User	Who are the users? What role do the users play? Who influences them?

EXPERT TIP
Hook framework

The hook framework (Alex Cowan) is based on the idea that a digital service or a product can become a habit for a user. The hook canvas is based on four main components: trigger for an action, activity, reward, and investment. For the potential user, there are two triggers for his actions: triggers from the external environment (e.g., a notification from Tinder that you received a "Super Like") or internal triggers for an action (e.g., visiting the Facebook app when you feel lonely).

The action describes the minimum interaction of your service or your product with a potential user. As a good designer, you want to design an action to be as simple and fast as possible for the user.

Reward is the key emotional element for the user. Depending on the configuration of the action, the user can be given a lot more than the satisfaction of the initial need. Think of positive reviews and feedback through a comment or article. You just wanted to share the information, but you get back far more due to the reputation of the community.

The question remains as to what the user invests in order to get himself back in the loop and to trigger an internal or external action. For example, he actively follows a Twitter feed or writes a notification that a certain product or service is available again.

The hook canvas

Trigger

1 External trigger

- What are the relevant triggers of an action for your various personas?
- What are the external and internal triggers to use your product or your service?

2 Internal trigger

- What does the user want, and how can we make him more effective?
- What existing triggers for an action are generally valid?
- How can we replicate the actions of the user?

Action

3 What is the simplest action that our user must perform to be rewarded?

Have we already minimized the effort to such an extent in order to perform the action for the user?

Investment

5
- How does our persona release the next action (investment of knowledge or development of a preference for a specific action)?
- What possibilities are there to close this loop in a better way?

Variable reward

4
- How is the user rewarded?
- Does the reward develop beyond the original goal?
- Is there a reward for the community and the potential user?

EXPERT TIP
Jobs-to-be-done framework

What is the actual task of a product?

The jobs-to-be-done framework became widely known through the milkshake example. The problem statement looks familiar to us: How can the sales of milkshakes be increased by 15%? With a conventional mindset, you would look at the properties of the product and then consider whether a different topping, another flavor, or a different cup size might solve the problem. Through a customer survey, you find out that the new properties are popular. However, in the end, only incremental innovations are realized, and the result has only been marginally improved. The jobs-to-be-done framework focuses instead on a change of behavior and on customer needs. In the case of the milkshake, it was found this way that two types of customers buy milkshakes in a fast food restaurant. The point of departure was: Why do customers buy a product? To put it differently: What product would they buy instead of the well-known milkshake?

The result:

The first type of customer comes in the morning, commutes to work by car, and buys a milkshake as a substitute for breakfast and as a diversion while driving. Coffee doesn't work because it is first too hot and then too cold. It is also liquid and can spill easily. The ideal milkshake is large, nutritious, and thick. So the jobs-to-be-done of the milkshake are therefore a breakfast substitute and a pleasant diversion while driving to work.

The second type of customer comes in the afternoon, usually, a mother with a child. The child wants something to eat in the fast food restaurant and is whining. The mother wants to get something healthy for the child and buys a milkshake. The milkshake should be small, thin, and liquid, so the child can drink it quickly, and it should

be low in calories. The milkshake's jobs-to-be-done are to satisfy the child and make the mother feel good. In principle, for any product, whether digital or physical, you can ask: Why would a customer buy my product or service?

Innovations like those designed by Adobe Photoshop and Instagram are good examples of jobs-to-be-done in the digital environment. Both solutions aim at making photographs look like those taken by pros. Photoshop offers easy professional editing of pictures through an app. Instagram realized early on that pictures can be easily edited and shared via social media.

Jobs-to-be-done, digital

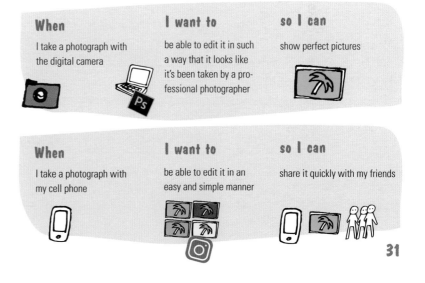

When	I want to	so I can
I take a photograph with the digital camera	be able to edit it in such a way that it looks like it's been taken by a professional photographer	show perfect pictures
I take a photograph with my cell phone	be able to edit it in an easy and simple manner	share it quickly with my friends

HOW MIGHT WE... develop a persona?

Because human beings always take center stage in design thinking and the persona to be created is very important, we sketched out the approach once more by way of example. When teams are tasked with developing "empathy" with a user over a certain period of time, or when they first apply design thinking, it is useful to specify a structure and the steps to be taken. Depending on the situation, we recommend using the tools just described (AEIOU, jobs-to-be-done framework, hook canvas, user profile canvas) or integrating and using other methods and documents into the steps listed here.

To help you better understand this process, the *Playbook* is interspersed with various "How might we . . ." procedures.

1. Find the user

Questions
Who are the users?
How many are there?
What do they do?

Methods
Quantitative collection of data,
AEIOU method

2. Building up a hypothesis

Question
What are the differences between the users?

Methods
Description of the groups of similar users/segmentations

10. Continuous further development

Questions
Is there any new information? Does the persona have to be newly described?

Methods
Usability test, continuous revision of the persona

9. Creating scenarios

Questions
In a given situation and with a given objective: What happens when the persona uses the technology?

Methods
Narrative scenario—storytelling, descriptions of situations, and stories in order to create scenarios
Application of hook canvas

3. Confirmations

Question
Is there any data or evidence that confirms the hypothesis?

Methods
Quantitative collection of data, empathy map

4. Finding patterns

Questions
Are the initial descriptions of the groups still correct? Are there other groups that might be important?

Methods
Categorization, applying the jobs-to-be-done framework

5. Creating personas

Question
How can the persona be described?

Methods
Categorization, persona

8. Dissemination of knowledge

Question
How can we present the personas and share them with other team members, the enterprise, or stakeholders?

Methods
Posters, meetings, e-mails, campaigns, events, videos, photos

7. Validation

Question
Do you know such a person?

Methods
Interviews with people who know the personas
Reading and commenting on the persona description

6. Define situations

Questions
What use cases does the persona have?
What is the situation?

Methods
Searching for situations and needs
User profile canvas/customer profile
Customer journey

EXPERT TIP
Future user

Especially in radical innovation projects, the time horizon is often far longer. It may take 10 years before a product is launched on the market, for example. If its target group is 30 to 40 years old, this means that these users now are 20 to 30 years old.

The future user method attempts to extrapolate these users' future personas (see "Playbook for Strategic Foresight and Innovation"). It expands the classic persona by analyzing today's persona and its development over the last few years. In addition, the future target group is interviewed at their present age. Subsequently, the mindset, motivation, lifestyle, etc. are extrapolated to get a better idea of the future user.

The method is easy to apply. It is best to start with the profile of the current user and underpin it with facts, market analyses, online surveys, personal interviews, and so forth.

When developing the persona, changes in values, lifestyle, use of technologies/media, product habits, and the like, must be borne in mind.

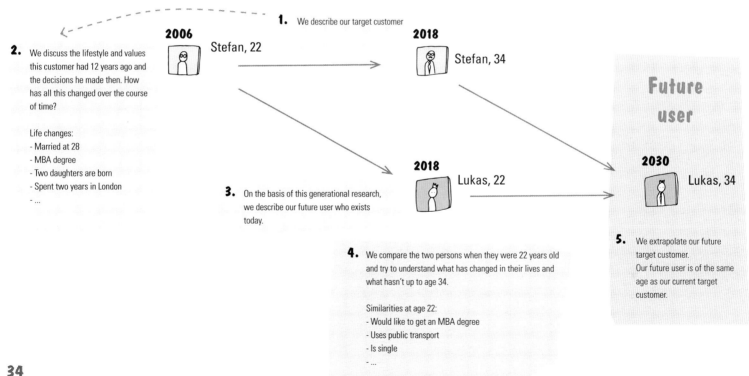

1. We describe our target customer

2006 Stefan, 22

2018 Stefan, 34

2. We discuss the lifestyle and values this customer had 12 years ago and the decisions he made then. How has all this changed over the course of time?

Life changes:
- Married at 28
- MBA degree
- Two daughters are born
- Spent two years in London
- ...

3. On the basis of this generational research, we describe our future user who exists today.

2018 Lukas, 22

4. We compare the two persons when they were 22 years old and try to understand what has changed in their lives and what hasn't up to age 34.

Similarities at age 22:
- Would like to get an MBA degree
- Uses public transport
- Is single
- ...

Future user

2030 Lukas, 34

5. We extrapolate our future target customer. Our future user is of the same age as our current target customer.

KEY LEARNINGS
Working with personas

- Use real people with real names and real properties.
- Be specific in terms of age and marital status. Get demographic information from the Internet.
- Draw the persona, in life-size, if possible.
- Add visualizations to the persona. Use clip outs from magazines for accessories (e.g. watch, car, jewelry).
- Identify and describe use cases in which they would use the potential product or services.
- Put the potential user in the context of the idea, his team, and the application.
- List pains and gains of the persona.
- Capture the customer tasks (jobs-to-be-done) that the product or service supports.
- Describe the experience that is particularly critical. Build a prototype that makes it possible to find out what is really critical.
- In so doing, try to take the persona's habits into account.
- Try out tools for the definition of the content (e.g., user canvas and customer profile, hook canvas, future user, etc).

An important factor of success in design thinking is to know where you stand in the process. For Lilly, Peter, and even Marc, the transition from a divergent to a convergent phase is a special challenge:

At what point in time have we gathered sufficient information, and how many ideas are necessary before we begin to transform the cavalcade of ideas into possible solutions?

Alongside the current level of development, the tools must be constantly kept in mind in design thinking. Which of them are the most effective in the current situation? There are generally two mental states in the "hunt for the next big opportunity": Either we develop many new ideas (i.e., we "diverge,") or we focus on and limit ourselves to individual needs, functionalities, or potential solutions (i.e., we "converge"). This is usually depicted in the shape of a double diamond.

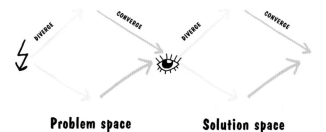

Problem space **Solution space**

For Lilly, it is a little easier to meet this challenge, because she knows how long her design thinking course at the university lasts, and she can control, as early as with the definition of the design challenge, how open or restrictive the question should be (i.e., how broad the creative framework for the participants is to be). With regard to real problem statements, things are somewhat different. Normally, we force ourselves at the beginning to leave our comfort zone and define the creative framework broader than we actually wish to. In the divergent phase, the number of ideas is infinite, so to speak. The tricky part here is to wrap up this phase at the right time and focus on the most important functionalities that ultimately lead to an optimal user solution. Of course, there are many examples of all sorts of ideas being launched on the market, and chance contributes to success—well-known examples include a number of services offered by Twitter. But chance does not often work this way; hence, in the process, converging is decisive for success.

Steve Jobs was a master when it came to managing the "groan zone" optimally. He had the right instinct to choose the time for a change of mindset and for leaving the divergent phase. This way, he led his teams to brilliant solutions. At Apple, Bud Tribble established the term "reality distortion field," standing for Steve Jobs's ability to master the mental leap. The term stems from an episode of the original *Star Trek* series, "The Menagerie," in which aliens create their own world by means of their thoughts.

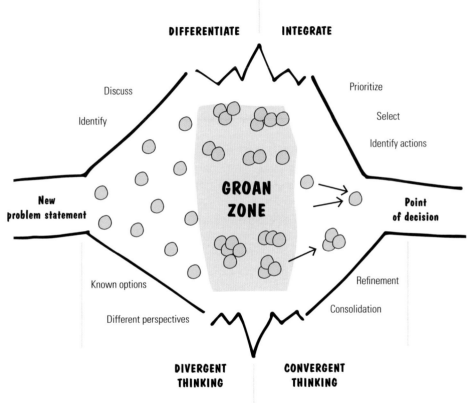

DIFFERENTIATE   INTEGRATE

Discuss   Prioritize

Identify   Select

                           Identify actions

**GROAN
ZONE**

**New
problem statement**   **Point
of decision**

Known options   Refinement

Different perspectives   Consolidation

**DIVERGENT   CONVERGENT
THINKING   THINKING**

One good influencer that helps us change our mindset is a limited period of time. If the final deadline for an innovation project is pushed forward or the first prototype is expected earlier than scheduled, the mindset must be automatically changed as well. In addition, it is advisable to lay down the functionalities and characteristics in an early phase of the design thinking process. During the transition to the convergent phase, we take them up again and attempt to match them to a great number of varying ideas. This selection makes it possible to eliminate some ideas at this stage. It can be useful for consolidating or combining ideas into logical clusters. But, even then, we won't be spared from selecting and focusing in the end. It is helpful during this phase to present the remaining ideas to other groups and participants. Then Post-its can be distributed, and the community can decide which is the best idea. If we only involve our own group, the decision is often not objective enough because we always risk falling in love with a certain idea. It's up to you how ideas that have not entered the convergent phase ought to be dealt with. Some facilitators encourage participants to throw the ideas written on the Post-its on the ground, while others keep the ideas as a knowledge reservoir until the end of the project.

What does the design thinking micro cycle look like?

Before we deal with the process in greater depth, we need to clarify the various design thinking processes, which basically all pursue the same goal but use different terms. Basically, there's a problem statement at the beginning and a solution at the end, and the solution is reached in an iterative procedure. The focus is decidedly on the human being, so design thinking is often referred to as Human-Centered-Design. Most people who have already grappled with design thinking know the process. Nonetheless, we decided to address briefly the phases in the micro cycle and the macro cycle as well as the core idea of each phase. Lilly would probably identify with the six-step depiction used at the HPI (Hasso Plattner Institute) that presents, as most universities do, the process of design thinking that follows. Subsequently, we will discuss the macro cycle.

At some universities, the process was simplified still more. In Japan, for instance, at the chair for Global Information Technology at Kanazawa Technical College, they work with four instead of six phases: Empathy—Analysis—Prototype—Co-Creation. D.school consolidates the process steps of "Understand" and "Observe" into "Develop empathy."

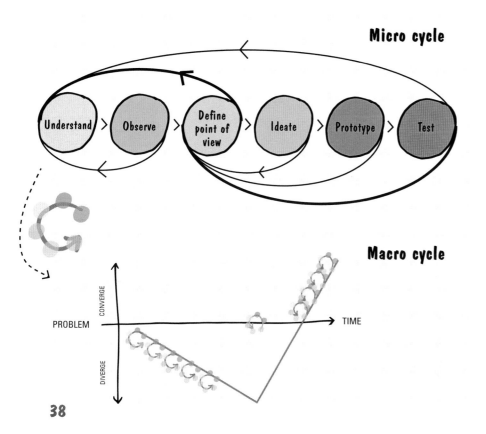

Micro cycle

Macro cycle

The IDEO design and innovation agency had originally defined five simple steps in the micro cycle in order to get to new ideas through iterations. In addition, they put a strong focus on implementation, because the best ideas are ultimately of no use if we haven't established them on the market as a successful innovation:

UNDERSTAND the task, the market, the clients, the technology, the limiting conditions, restrictions, and optimization criteria.

OBSERVE and ANALYZE the behavior of real people in real situations and in relation to the specific task.

VISUALIZE the first solution drafts (3D, simulation, prototypes, graphics, drawings, etc.).

EVALUATE and OPTIMIZE the prototypes in a fast succession of continuous repetitions.

IMPLEMENT the new concept in reality (the most time-consuming phase).

Scrum process

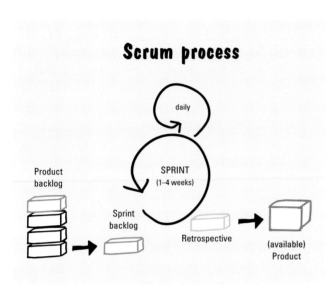

Product backlog
Sprint backlog
daily
SPRINT
(1–4 weeks)
Retrospective
(available) Product

Hear --→ Create --→ Deliver

Anybody working at an actual business ought to know iterative procedures in a different context, such as from software development (ISO Standard 13407 or Scrum). In this case, the user suitability of software is ensured by an iterative process or improved incrementally through sprints.

In ISO 13407, the following phases are spoken of:

Planning, process—Analysis, use context—Specifications, user requirements—Prototype (draft of design variants)—Evaluation (evaluation of solutions and requirements)

With Scrum, the individual iterations are called *sprints*. One sprint takes 1 to 4 weeks. So-called product backlogs serve as inputs into the sprints. They are then prioritized and processed in the sprints (sprint backlogs). The requirements are documented in the form of user stories in the product backlog. A ready-to-deliver product that has already been tested with the user during the sprint is what should exist at the end of each sprint. In addition, the process itself is reviewed and continually improved in the Retrospective.

At most companies, a micro design thinking process is broken down into three to seven phases, often based on the steps of IDEO, d.school, and the HPI. The Swiss ICT company Swisscom has designed a simplified micro cycle that allows for integrating the mindset quickly into the organization.
The phases are: **Hear—Create—Deliver.**

Phase	Description	Basic tools
Hear	- Understand the project - Understand the customer problem/need - Procure information, internal and external - Gather experience directly from the customer	Design challenge Customer interview
Create	- Transform what was learned into potential solutions - Generate multiple solutions and possibilities - Define solution features	Core beliefs Target customer experience chain
Deliver	- Concretize ideas - Create and test prototypes - Verify, expedite, or reject ideas - Gain insights and learn from them	Need, Approach, Benefit, Competition (NABC) Prototyping plan Self-validation

EXPERT TIP
The design thinking micro cycle

DESCRIPTION OF THE INDIVIDUAL PHASES OF THE MICRO CYCLE

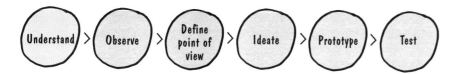

Understand > Observe > Define point of view > Ideate > Prototype > Test

UNDERSTAND:

This phase was already touched upon in Chapter 1.1. Our starting point was not a goal to be achieved but a persona that has needs or is facing the challenge of having to solve a problem. Once the problem has been recognized, the problem statement must be defined at the right level of comfort. With two types of questions, we can either expand (WHY?) or narrow down (HOW?) the creative framework. The principle can be illustrated most easily on the basis of the need to educate ourselves further:

Alongside the problem statement, it is important to understand the overall context. Answering the six WH questions (who, why, what, when, where, how) yields fundamental insights:

- Who is the target group (size, type, characteristics)?
- Why does the user think he needs a solution?
- What does the user propose as a solution?
- When and for how long is the result needed (time span of the project or life cycle of the product)?
- Where is the result going to be used (environment, media, location, country)?
- How is the solution implemented (skills, budget, business model, go-to market)?

More on this in Chapters 1.4 and 1.5.

TAKE CARE OF OURSELVES AND OUR FAMILIES

ACHIEVE INDIVIDUAL FULFILLMENT

INCREASE THE ECONOMIC VALUE FOR OUR SOCIETY

WHY?

EDUCATE OURSELVES FURTHER IN DESIGN THINKING

ARCHIVING INFORMATION

HOW?

READING *THE DESIGN THINKING PLAYBOOK*

LEARNING FROM OTHERS

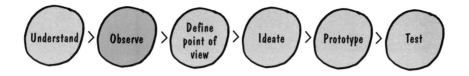

OBSERVE:

We have already initially dealt with the Observe phase to some extent. We tried to be experts and better understand the needs of our readers. We took a closer look at people from three different environments who apply design thinking and observed the groups of persons at work. To do so, we took advantage of various opportunities: at the HPI in Potsdam, at the d.school in Stanford, interacting with coaches from the ME310; in workshops with the DTP Community at Startup Challenges; in internal workshops at companies as well as in co-creation workshops with the objective of inspiring customers for digitization; and so on.
It is always important to document and visualize these findings so they can be shared with others at a later time. So far, most of those involved in design thinking focus on the qualitative method of observation. Documentation is done by means of idea boards, vision boards, daily story based on photos, mind maps, mood pictures, and photos of life situations and people. All this is important information we can use to create and revise personas and to build up empathy for the user, as will be described in more detail in Chapter 1.5.

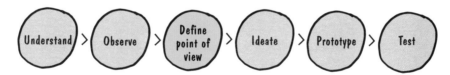

DEFINE POINT OF VIEW:

For the point of view, the important thing is to draw upon, interpret, and weight all the findings. The facilitator is urged to encourage all members of a group to talk about their experience. The goal is to establish a common knowledge base. This is done best by telling stories that have been experienced, showing pictures and describing the reactions and emotions of people. Again, the aim is to develop further or revise the personas in question. We will discuss this step in detail in Chapter 1.6.

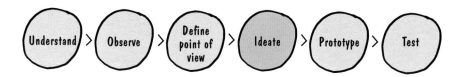

IDEATE:

In the phase of Ideation, we can apply various methods and approaches that heighten creativity. Irrespective of this, we normally use brainstorming or the creation of sketches in this phase. The goal is to develop as many different concepts as possible and visualize them. We present a number of techniques for this in Chapter 1.7. The phase of Ideation is closely associated with the subsequent phases in which prototypes are built and tested. The next Expert Tip will give depth to this approach. In this phase, our primary goal is the step-by-step increase of creativity per iteration. Depending on the problem statement, a general brainstorming session on possible ideas can be held at the onset. Presenting individual tasks in a targeted manner for the brainstorming session has proven successful; this way, creativity and thus the entire diverging phase can be controlled. Examples include a brainstorming session on the critical functions, benchmarking with other industries or situations, and a dark horse that deliberately omits the actual situation or combines the best and worst ideas. A funky prototype that simply ignores all limiting factors can also generate ideas. We will specifically address the matter in the depiction of the macro cycle.

PROTOTYPE:

In the previous phase, we already pointed out the next steps of "Build prototype" and "Test prototype" because they are always connected to ideation. Chapter 1.9 will show what makes up a prototype.

At any rate, we should make our ideas tangible as early as possible and test them with potential users. This way, we receive important feedback for the improvement of ideas and prototypes. The motto of the options for action is simple: Love it, change it, or leave it.

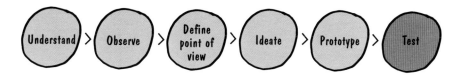

TEST:

This phase comes after each developed prototype and/or after each drafted sketch. We can do the testing with colleagues, but the interaction with potential users is what's really intriguing. Alongside traditional testing, it is possible today to use digital solutions for testing. Prototypes or individual functionalities can be tested quickly and with a large number of users. We will present these possibilities in Chapter 1.10. We receive mostly qualitative feedback from this phase. We should learn from these ideas and develop them further until we love our idea. Otherwise: discard or change.

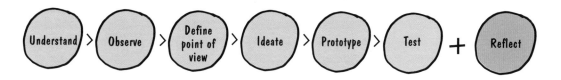

REFLECT:

Before starting a new cycle of the iterative process, it is worthwhile to reflect upon the chosen direction. Reflection is best triggered by questioning whether the ideas and test results comply with claims of being socially acceptable and resource-efficient. With agile methods such as Scrum, the Reflect phase wraps up the process in retrospection. The process and the last iteration are reviewed, and a discussion follows on what went well and what should be improved. The questions can be played through in a "I like—I wish" feedback cycle, or feedback can be obtained in a structured way using a feedback capture grid. Naturally, we also use the Reflection phase to consolidate the findings if this hasn't yet occurred in the Test phase.

We update the personas and, if necessary, other documents on the basis of these findings. In general, reflecting helps to explore new possibilities that might lead to better solutions or improve the process as a whole.

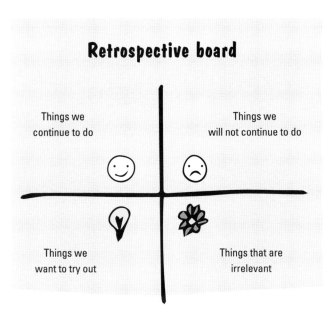

Retrospective board

Things we continue to do

Things we will not continue to do

Things we want to try out

Things that are irrelevant

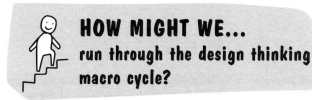

In the micro cycle, we go through the phases of Understand, Observe, Define point of view, Find ideas, Develop prototype, and Test proto- type. They must be seen as a unit. In the divergent phase, the number of ideas we gather through various creativity techniques increases constantly. Some of these ideas we want to make tangible in the form of prototypes and test with a potential user. The respective creativity methods and tools are used depending on the situation. **The journey toward the ultimate solution is not certain at the onset.**

The issue in the macro cycle is to understand the problem and concretize a vision of the solution. To do this, many iterations of the micro cycle are run through. The initial steps in the macro cycle are of a diver- gent character (steps 1–5 in the figure). In the case of simple problems or if the team possesses compre- hensive knowledge of the market and the problem, the transition to the groan zone (step 6) can be pretty fast. Transition to the groan zone can be effected from any one of the five divergent steps. The sequence of ideas to be elaborated can and must be adapted to the situation and the project. The suggested sequence has been successfully applied in many projects, though. The vision of the solution or idea is concretized in the form of a vision prototype and tested with different users. If the vision gets generally positive feed- back, it is concretized in the next iteration (step 7).

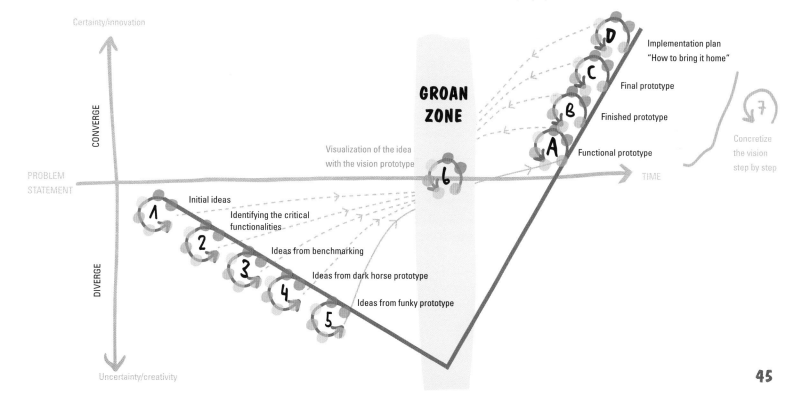

The hunt for the next big market opportunity often follows these steps:

(1) Initial ideas are worked out in a brainstorming session

An initial brainstorming session about potential ideas and solutions helps the group to place all sorts of ideas and get them off their collective chest. Frequently, the levels of knowledge of the individual team members in terms of the problem statement and a possible solution spectrum are quite different. An initial brainstorming session helps in approaching the task and learning how the others in the group think.

Instruction: Give the group 20 minutes for a brainstorming session. The issue here is quantity, not quality. Every idea is written on a Post-it. When writing or sketching on the Post-it, the idea is expressed aloud; afterward, the note is stuck to a pin board.

Ask the group to answer the following key questions:
- Which ideas come to mind spontaneously?
- Which solution approaches are pursued by the others?
- What can we do differently?
- Do we all have the same understanding of the problem statement?

Brainstorming

(2) Develop critical functionalities that are essential for the user

This step can be crucial for the solution. The facilitator has the task of motivating the groups so they identify exactly these "important things" and prepare a ranking in the context of a critical user.

Instruction: Give the group one to two hours—depending on the problem statement—to draft, build, and test 10 to 20 critical functions.

Ask the group to answer the following key questions:
- Which functionalities are mandatory?
- What experience is absolutely necessary for the user?
- What is the relationship between the function and the experience?

Critical functions

(3) Find benchmarks from other industries and experiences

This step is a very good tool when teams are not able to tear themselves away from an original solution concept.
Benchmarking helps participants think outside the box and adapt ideas from these areas for the solution of the problem. The facilitator broadens the creative framework by motivating the groups to hold the brainstorming session, taking into account a certain industry/sector or a particular experience. You can proceed in two steps, for instance: (a) brainstorming of ideas relating to the problem, and (b) brainstorming of industries and/or experiences. Subsequently, the three best ideas from each step are identified. Based on the combination of these, the facilitator invites the participants to develop two or three ideas further, build them physically, and test them with the user.

Instruction: Give the group 30 minutes for a brainstorming session, 30 minutes for finding benchmarks, and 30 minutes for clustering and combining ideas. Depending on the task, the group is given enough time to build two to three prototypes.

Ask the group to answer the following key questions:
- Which successful concepts and experiences can be applied to the problem?
- Which experiences can illuminate the problem from another perspective?
- What is the relation between the problem and other experiences?

Benchmarking

(4) Heighten creativity and find the dark horse among the ideas

This **step helps many teams to boost creativity further—not least because, for the dark horse, borders are lifted**, which might have limited us in the previous steps. The facilitator motivates the groups to strive for maximum success and thus develop a radical idea. Now the time has come for the teams to heighten creativity and accept the maximum risk. One possibility for the creation of a dark horse is to omit essential elements of a given situation, such as, "How would you design an IT service desk without IT problems?" "What does a windshield wiper look like without a windshield?" or "What would a cemetery look like if no one died?" The main point is to leave the comfort zone and "do it in any case," no matter what will occur.

Instruction: Give the group 50 minutes to build a dark horse and enough time for building a corresponding prototype depending on the task.

Ask the group to answer the following key questions:
- Which radical possibilities have not been considered thus far?
- Which experiences lie outside anything imaginable?
- Are there products and services that would expand value creation?

DARK HORSE **What if?**

(5) Implementation of a funky prototype to give free rein to creativity

In many cases, you have to go one step further because the team has not come up with disruptive ideas so far. The building of a funky prototype cranks up creativity one more notch. It encourages the teams to maximize the learning success and at the same time minimize costs in terms of time and attention. The goal is to develop solutions that mainly focus on the benefit. Potential costs and any budget restrictions are completely removed.

Instruction: Give the group an hour to build a funky prototype.

Ask the group to answer the following key questions:
- What crazy ideas are super cool?
- For which idea would you have to ask forgiveness in the end?
- What does an idea look like that is realized ad hoc and has not been planned?

(6) Determine the vision of the idea with the vision prototype

The groan zone is the transition from the convergent to the divergent phase. The phases can be changed at any time. Experienced facilitators and innovation champions recognize this point in time and lead their teams in a targeted way to the convergent phase. In the vision prototype, we make an initial combination of

- prior knowledge (caution is advisable here),
- best initial ideas,
- most important critical functionalities,
- new ideas of other industries and experience,
- initial user experience,
- intriguing insights (e.g., from the dark horse), and
- the simplest possible solution.

Instruction: Give the group about two hours (depending on the complexity of the problem) for building a vision prototype. It should then be tested with at least three potential users; the feedback is to be captured in detail. In the best-case scenario, these users are then involved in the subsequent concretization of the design thinking project. If so-called lead users are known in a field of innovation, they are perfect as references because they are often highly motivated to satisfy their needs.

Ask the group to answer the following key questions:
- Does the vision generate enough attention so a potential user absolutely wants to use this solution?
- Does the vision give sufficient leeway for a user's dreams?
- Is the value offer of the vision convincing?
- What else would the users wish for in order to make the experience perfect?

$$1+1 = 3$$

(7) Concretize the vision step by step

In the following convergent phase, we want to focus on the concretization of the vision.
The theme of this phase is the specific elaboration of the selected idea. It is iteratively improved and expanded. It is advisable here first to build and test the most important critical functionalities as integral parts of a functional prototype. With this prototype as a starting point, more elements are supplemented and finally the prototype is built. Different ideas can be tested in the convergent phase, and the best ones are integrated into the ultimate solution. Individual features or various combinations can be developed and tested, for instance. Once the prototype has a certain maturity, we can describe it in a "prototype vision canvas." This way, we can formulate and compare various visions.

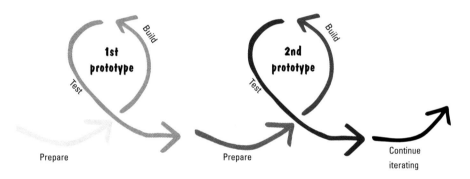

Prototype vision canvas

Vision statement:			
Target group:	Needs:	Product:	Benefit:

It is all about the iterative detailing and elaboration of the selected idea.
The maturity of the prototypes increases with every individual step.

A. Functional prototype

With respect to the functional prototype, it is important to concentrate on the critical variables and test them intensively with potential users. Critical functions must be created for critical experiences. Not all functionalities must be integrated at the onset. The crucial point is to ensure minimum functionality in order to test the prototype under real conditions. These prototypes are frequently referred to as "minimal viable product" (MVP). These MVPs serve as a foundation to build upon, and step by step a finished prototype emerges that combines several functions.

B. Finished prototype

The creation of a finished prototype is crucial for the interaction with the user, because only reality yields truth. Enough time must be scheduled for building a finished prototype, and the respective functionalities must be integrated.

C. Final prototype

The final prototype excels by the elegance of the thoughts invested in it as well as in its realization. Prototypes that are convincing with simple functionality are usually also successful when launched on the market. It is advisable to obtain as much support from suppliers and partners in any and every possible way. The use of standard components increases the likelihood of success and massively reduces development costs.

D. Implementation plan: How to bring it home

Not only the quality of the product or service is decisive, but also its implementation. Important things to know: Who might put obstacles in the way of the implementation process and try to influence decisions? The credo: Turn those affected into people who are involved and create a win-win situation for all parties. Chapter 3.4 describes what is important in the implementation process.

KEY LEARNINGS
Keeping a grip on the process

- Define a problem statement on the right level.
- Leave the comfort zone (as often as possible) if you want radical innovations to emerge.
- Develop awareness of the groan zone in the macro cycle, because it is decisive for the future success of the generated ideas.
- Create clarity on the team about whether the divergent or convergent mindset is currently at center stage.
- Use different methods in the divergent phase for brainstorming in order to heighten creativity (e.g., benchmarking, funky prototype, dark horse).
- Generate as many ideas as possible in the divergent phase by applying different creativity techniques.
- Always follow the sequence of "Design—Build—Test" in the micro cycle.
- Find the final prototype through converging and the respective iterations.
- Don't develop emotional ties to prototypes and ideas. Discard bad ideas.
- What applies to all ideas: Love it, change it, or leave it!

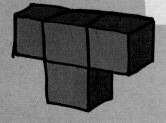

1.3 How to get a good problem statement

At the beginning, Peter didn't understand why it's important to have a good problem definition in design thinking. After all, he wanted to find good solutions and not make the problems worse. During his first tentative steps as a facilitator for design thinking workshops, though, he quickly noticed, just how important the problem definition is. He realized there are three essential prerequisites for good solutions:

1. The design thinking team must have understood the problem.
2. The design challenge must be defined to allow for the development of useful solutions.
3. The potential solution must fit the defined design space and design scope.

We break down problems into three types: simple (well-defined), poorly defined (ill-defined) and complex (wicked). For simple and clearly defined problems, there is one correct solution, but the solution strategy can follow different paths. Most problems we encounter in design thinking and in our daily work are ill-defined problems, however. They can be remedied with more than just one correct solution, and the search for such a solution can take place in quite different ways. From our experience, we nevertheless know that these problems can be rendered graspable and easily processed. Often it's enough to reduce the creative framework or sometimes widen it a bit to get to the right level that allows new market opportunities to emerge.

Repeatedly asking "Why?" expands the creative framework; asking "How?" scales it down. In the Introduction, we briefly referred to the question of how we would tackle the issue of further training with design thinking. Designing a better can opener that everybody in the family likes using is another simple example of a design challenge.

To expand the problem statement, we pose the question of "Why?". Quickly we realize that repeatedly asking why brings us to the limits of our comfort zone in no time at all, so that we are actually moving toward earth-shaking and difficult-to-solve problems, so-called wicked problems. In terms of the can opener, examples of such problems are:

- How can we stop hunger in the world?
- How can we prevent so much food from being thrown away?

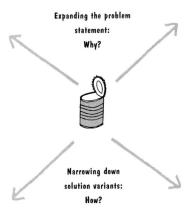

Expanding the problem statement: Why?

Narrowing down solution variants: How?

To narrow down alternative solutions, it helps to ask "How?" With regard to the can opener:
- How can the can be opened with a rotating mechanism? or:
- How can the can be opened without any additional device?

PROBLEM

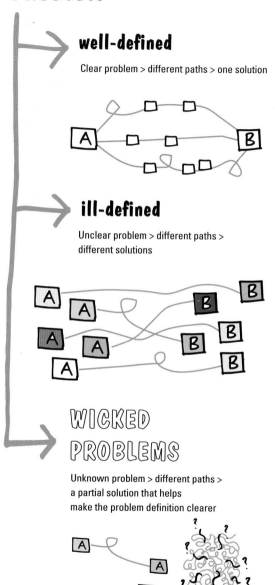

well-defined

Clear problem > different paths > one solution

ill-defined

Unclear problem > different paths > different solutions

WICKED PROBLEMS

Unknown problem > different paths > a partial solution that helps make the problem definition clearer

Regarding wicked problems, the actual issue is often not obvious, so preliminary problem definitions are used. This leads to an understanding of the solution that changes the understanding of the problem again. So there are iterations already in the problem definition that can help interpret the understanding of the problem as well as of the solution. Only short-term or provisional solutions are largely found by way of this co-evolution, though. The use of linear and analytical problem-solving procedures quickly makes you hit your limits in terms of wicked problems: Because the problem is the search for the problem, you're pulled every which way.

Fortunately, relevant tools for this were discovered in design thinking over the years, such as the question of "How might we . . .?" or a technique regarding "why" questions. Thus design thinking helps to make wicked problems graspable. If no solutions are found despite the use of design thinking due to the complexity of the problems, limited resources such as money and time are usually the reasons for the termination of the process. This is why we recommend devoting enough time and energy to work out the definition of a suitable problem definition.

To which types of problems can design thinking be applied?

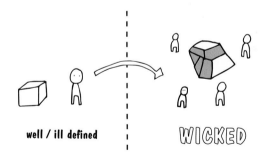

well / ill defined   WICKED

Design thinking is suitable for all types of problem statements. Applications range from products and services to processes and individual functions, all the way to comprehensive customer experiences. But the goals people want to achieve with it differ. A product designer wants to satisfy customer needs, while an engineer is more interested in defining the specifications.

In her design thinking courses, Lilly often has difficulty finding good design challenges. If the design challenges come from industry partners, the creative framework is usually more or less set.

In cases where the participants must identify problems on their own, things get more complex. The following options have proven quite useful for identifying problems and defining design challenges:

How might we improve the customer-experience chain of places and things that are visited or used daily?

Examples:

- How might we improve the online shopping experience of a shoe retailer?
- How might we improve the online booking portal for the car ferry from A to B?
- How might we improve customer satisfaction with the ticket app for public transport in Singapore?

Design the **perfect customer experience** for the online booking of a car ferry

Another possibility for getting to a design challenge is to change perspective. These questions help capture the design challenge:

- What if . . .?
- What might be possible?
- What would change behavior?
- What would be an offer if business ecosystems connected with each other?
- What is the impact of a promotion?
- What will happen afterward?
- Are there any opportunities where other people only see problems?

Another possibility is to take a closer look at an existing product or service (e.g., the customer experience chain when buying a music subscription). By asking questions and observing, we get hints for a design challenge:

- What does the music behavior of a user look like?
- How does the customer get information on new music offerings?
- How and where will the customer install the product or service?
- How does the customer use the product?
- How does the customer act when the product does not work as expected?
- How satisfied is the customer with the entire customer experience chain?

The description of the design challenge is definitive. As we remarked, a good solution can only come about if the design thinking team has understood the problem.

The description of the challenge must be seen as a minimum requirement. Further details help to expedite the problem solving. The disadvantage here is that the degree of freedom in relation to the radicalism of a new solution is limited. The creation of a good design brief (short profile of the project) is already a small design thinking project in itself. Sometimes, we draw up the design brief for our users, sometimes for the design thinking team. We recommend you get different opinions—preferably on an interdisciplinary basis—about the problem and then agree, through iterations, on statements that really make up the problem.

The design brief contains various elements and can provide information on core questions:

Definition of design space and design scope:
- Which activities are to be supported and for whom?
- What do we want to learn about the user?

Description of already existing approaches to solving the problem:
- What already exists, and how can elements of it help with our own solution?
- What is missing in existing solutions?

Definition of the design principles:
- What are important hints for the team (e.g., at which point more creativity is demanded or that potential users should really try out a certain feature)?
- Are there any limitations, and which core functions are essential?
- Whom do we want to involve, and at what point in the design process?

Definition of scenarios that are associated with the solution:
- What does a desirable future and vision look like?
- Which scenarios are plausible and possible?

Definition of the next steps and milestones:
- By when should a solution have been worked out?
- Are there steering committee meetings from which we can get valuable feedback?

Information on potential implementation challenges:
- Who must be involved at an early stage?
- What is the culture like for dealing with radical solution proposals, and how great is its willingness to take risks?

A design brief is the translation of a problem into a structured task:

DESIGN BRIEF:

Why?	>	Problem
Who?	>	Target customer
What?	>	Goals
With what?	>	Available material
Who else?	>	Competition, alternatives
How much?	>	Budget, restrictions
When?	>	Schedule, scenarios
How?	>	Next steps

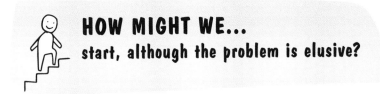

HOW MIGHT WE...
start, although the problem is elusive?

In principle, the ideal starting point is where we leave the comfort zone. To find the right starting point based on a problem statement is not very easy. Often the team wonders whether the starting point is too narrow or too broad. In such a case, we recommend just starting. If the challenge is too narrowly conceived, the team will expand the problem in the first iteration. If the challenge is conceived too broadly, the team will narrow it down.

Do we want to improve the cap of a ball-point pen or do we want to solve the world's water problem?

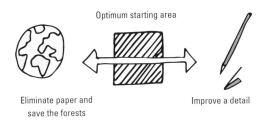

Optimum starting area

Eliminate paper and save the forests

Improve a detail

The procedure consists of three steps.

Step 1:
Who is the user in the context of the problem statement?
Define who the user really is and what his needs are.
Reflect on the created persona.

Step 2:
Apply the WH questions. Discuss the WHY, the WHAT, and the HOW.

Step 3:
Based on this, formulate your question.

Example

Step 1: Whom does it concern, and what is the central task?
Lilly holds innovation workshops as a facilitator.

For whom?

Step 2: What does Lilly want to achieve?

Lilly wants to document the information from the workshops.

What?

What does the user want to achieve?
Edit? Create? Evaluate?
Analyze?
Store? Share?

Step 3: Based on this, formulate your question

Lilly documents the question.

How can I digitally capture a Post-it and share it?

Why?

The results are to be
shared with others on the team
worldwide.

How?

Copy?
Speech recognition?
Text recognition?
Image recognition?
Photographing?

As described, the "why" and the "how" questions can expand or narrow down the framework.
A natural adjustment often takes place in a brainstorming session, especially if various methods are used in brainstorming, such as transforming and combining or even minimizing.

Method	How can we solve our problem?	
Minimize	reduce it?	reduce an existing solution?
Maximize	expand it?	expand an existing solution?
Transform	mentally transfer it to another area?	transfer a solution existing in another area to my problem?
Combine	combine it with other problems?	combine several existing solutions?
Modify/adapt	modify it?	modify an existing solution?
Rearrange/invert	change or invert its internal order?	change or invert the order of an existing solution?
Substitute	substitute a partial problem?	substitute a part of an existing solution?

Let's take the example of a BIC ballpoint pen and the method of leaving out or reducing. For the BIC ballpoint pen, everything that was unnecessary was left out. All that was left in the end were three indispensable, essential parts: the refill, the holder, and a cap that also serves as a clip. An ingenious product that has remained unaltered for more than 50 years.

Is there even any room left for innovation?

The answer is: Yes! Perhaps you have already asked yourself before why the cap of the BIC ballpoint pen has a hole at the tip. The hole was not always there. It was designed to prevent small children from suffocating if they swallow the cap and it gets stuck in their windpipe. Sufficient air can still get through the little hole. This is why BIC pen caps have had holes for more than 24 years.

KEY LEARNINGS
Draw up the problem definition

- Question in the form of "Why?" and "How might we?" in order to grasp and understand the problem.
- Clarify what type of a problem it is: wicked, ill-defined, or well-defined. Adjust your approach accordingly.
- In the case of wicked problems, first find partial solutions for a partial problem. Proceed iteratively.
- Understand further partial aspects of the overall problem if it can't be understood at once, and iteratively add more solution components.
- Draw up a structured design brief so that the team and the client have the same understanding of the starting point.
- Make use of different possibilities of finding the design challenge (e.g., investigation of the entire customer experience chain or a change of perspective).
- Begin with the first iteration even if the ideal starting area has not been found yet. This way, the problem can often be understood better.

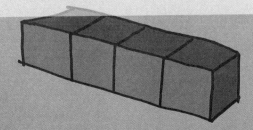

1.4 How to discover user needs

Priya has a new innovation project. Rumors have it that the Internet and technology giant where Priya is working will embrace the theme of health for seniors—a theme and a segment about which Priya knows little and which, for her personally, is still pretty remote. Actually, Priya has little time for taking the needs of seniors into consideration alongside her numerous other projects. Her work environment teems with people in their mid-twenties; hardly anyone has yet crossed the threshold of 50 and can be classed even remotely in this segment. Her friends and acquaintances in Zurich are all between 30 and 40 years old, and her parents are still working full time and don't feel they belong in the user group of retirees. Her grandparents, whom Priya could ask, have unfortunately passed away.

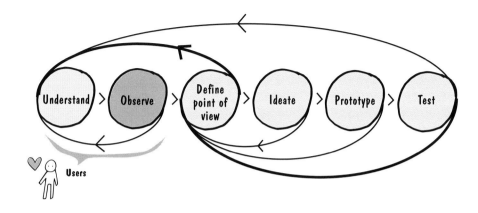

How can we carry out a needfinding when we actually have no time for it? Or better: How do we explain to the boss that we won't come to work today?

Priya is aware that the personal contact with potential users—that is, people—is indispensable if you really want to live good design thinking.

Omitting the needfinding is not an option for Priya, because it would mean skipping over an entire phase of the design thinking process. Because the phases of understanding and observing as well as the synthesis (defining the point of view) cannot be strictly separated from one another, ignoring needfinding would mean omitting no fewer than three steps.

All these steps have an important feature in common: the direct contact with the users, the target group of people who will use an innovative product or our service regularly in the future.

It is an illusion to think that we are familiar with the lifestyles of all the people for which we develop innovations day after day. Let's take a look at all the projects Lilly has gone through over the last four years as a needfinding expert: She would have had to be old, visually impaired, lesbian, a kindergartener, or even an illegal immigrant. Not to mention the project concerning a palliative care ward that inevitably would have catapulted Lilly into her deathbed. That certainly didn't happen to Lilly. At least not at the time when her task was to innovate everyday life for these people in the final hours of life and the procedures at a palliative care ward.

It is important to reflect on ourselves and realize we don't represent the people for whom we develop our innovation. If we do, in very exceptional cases, we must proceed with great caution when transferring our needs onto others.

Self-reflection

Peter also questions his ideas for improving the quality of his product when he is sitting at his desk, doing nothing. When was the last time he saw somebody using his product in daily life? Has he ever stood next to a customer at the exact moment when the customer felt the need for the now newly invented function? Not because Peter had asked the customer ("Would you like . . .") but because the customer had searched for this function on his own.

Such moments give us an insight into the lives of users and indicate where deep and long-term needs are hidden.

Not knowing the everyday life of people means we continually make assumptions on which we base our decisions. About eight million people live in Switzerland. If Priya, who lives in Zurich today, claimed she knows exactly how the residents in a small village live, then her knowledge is solely based on the experience of her youth when she lived in a village in India, that was about the same size at that time. Although her experience gives her access to certain aspects of village life, she is incapable of developing a perfect solution that covers the majority of needs of villagers in today's Switzerland.

It stands to reason that an innovation only works when we have internalized the needs of our users and developed a thorough understanding of them. It can be achieved when we are where they are, especially when we witness the part of their life we want to improve.

If you now think we'll present even more tools to observe people in their environment, you're wrong. Such tools can help us, but ultimately it is all a matter of one decisive point in needfinding: Find out which assumptions you have made in your mind and become aware of them.

In the everyday work of a company, it is a common phenomenon that innovation managers work on ideas that are not based on real needs. Often, when we ask them what doesn't work in the everyday life of a person that would give their ideas real added value, we are met with a blank stare.

In such cases, it is useless to send out the innovation managers, because they don't know what they should see and hear. So needfinding does not take place in many companies and is inevitably seen as a waste of time and money.

Many traditional management and innovation consultants rely on so-called customer interviews conducted not by the consultant himself or a market research institute tasked by him. The consultant then picks and chooses from the interviews only those things that match what he has seen or heard and that fit into the reality he has developed over a lifetime. Thus, not infrequently, decisions makers see needfinding as a risk to the success of their project.

If we succeed in embodying an attitude of pure curiosity in needfinding, we find that everything we learn can guide us to new and even more human-centered solutions.

In needfinding, we recognize things that still don't work, maybe that never will work, or that we must watch very closely so that, in the end, our innovation meets a need.

When was the last time we "walked in our customer's shoes."

When was the last time we mastered the daily grind at exactly the spot where our users are standing? If not for a whole day, then at least for an hour!

How do we know what our customers have difficulties with? What are the reasons that our customers are happy? What triggered the Wow! effect when they experienced our product?

There are a couple of good tricks that help free you of assumptions. Especially when dealing with needfinding for the first time, the following exercise, which doesn't take longer than 30 minutes, is highly recommended. When we have somebody who confronts us with tricky questions in this exercise, we will be all the more effective.

The purpose of this method line is to show how assumptions and hypotheses about needs can be made visible and how we succeed in prioritizing critical assumptions. This creates a starting point that enables us to realize a focused and, hence, more successful user interaction.

The starting point is that we have already built an initial simple prototype. Hence the phase of ideation has been concluded for now because we have already found a potential solution for a user need. Within the scope of her "health for seniors" project, Priya has identified the theme of exercise as an approach to a solution.

1. We formulate our idea in one sentence:
For example:
Senior walks for retired "couch potatoes."

Then we visualize our idea:

Senior wizard

2. We formulate the need assumptions of our idea:

As we know, needs are the actual motivations of people. They emerge from the desire to make something possible that does not exist (in our example: staying healthy) or to get rid of something not wanted (e.g., losing weight). In design thinking, we often define these needs as verbs. Needs refer to WHAT the user wants to achieve—we consciously put aside solution-oriented thinking, which is focused on the HOW.

To identify need assumptions, we first ask the following questions:

- What does the user want to achieve by applying our idea?
- What motivates the user to use our idea?
- What prevents the user from using our idea?

Possible answers include:

1. Couch potatoes want to exercise (need) in order to prevent chronic diseases (need).
2. Retirees don't have the necessary daily structure (trigger) to exercise on a regular basis (need).
3. Senior citizens want to feel healthy (need) so they can go on excursions with their grandchildren.
4. Senior citizens feel uncomfortable (emotional state/blocker) when they exercise at the fitness center together with young people.

Write each of these assumptions on a separate Post-it. Then you can place the Post-its on a grid in step 5.

3. We identify the critical assumptions:

First of all, it is important for us to take a few minutes to reflect upon our assumptions of needs.

What will we recognize in this phase of reflection? Perhaps we recognize we've dealt with the basic needs of our potential innovation—often, a wonderful crop of assumptions on which we have built our solution! Now these needs must be reviewed and adapted, if necessary.

With this exercise, we are confronted with the basis of our ideas without having heard or seen whether a potential user actually has a need for such an innovation in his everyday life.

Maybe we have found a couple of colleagues from among our friends who think our solution is good. Now it would be exciting for us to find out whether the parents and grandparents of our friends really have these problems in everyday life. With this step, we have gotten very close to our user. At the same time, we must be aware that we are still dealing with assumptions. We have not yet heard or seen whether these needs actually exist out there in real life.

We'll have no choice but to review these needs—this time, not with our work colleagues! We must observe and interview people who are not close to us and who won't react positively to our ideas because they like us or don't want to dampen our enthusiasm.

4. We are ready for random encounters:

What would we ask users in our target group if we met them by chance on the street now? In order to be prepared, we should seriously consider what question would we use to approach somebody to tell us about their everyday life. Priya, for example, ought to think about where and when she can meet retirees during the week in their everyday lives (e.g., shopping, on a trip, on the train, at the bus stop, etc.). The good news for Priya is that she doesn't have to take a single day off to conduct a needfinding. She can simply integrate it into her everyday life.

What is needfinding really about?

We must leave our comfort zone and speak to people in order to get a look at ideas from a new angle. We must be willing to learn new things and stay curious, enriching our knowledge step by step.

5. We review the critical assumptions first:

We should ask ourselves about which assumptions we know least and which are most critical for our idea. It's best we review these assumptions first.

If these assumptions do not exist in everyday life, we have built our solution idea on a mental castle in the air. This is not so bad, because the sooner we recognize it, the better it is for us. It saves a lot of money, time, and energy. We can use the freed-up resources to hunt for the next big market opportunity.

The review of the critical assumptions can be structured in the shape of four quadrants. Using the dimensions of "incidental" versus "decisive" and "knowing" versus "ignorant" has served us well in the past.

Mental castle in the air?

Can I help?

REVIEW CRITICAL ASSUMPTIONS

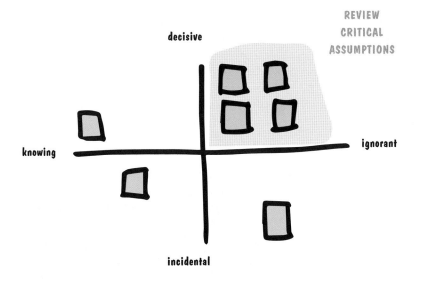

decisive

knowing

ignorant

incidental

HOW MIGHT WE...
conduct a needfinding interview?

Every interview should have a logical sequence. We recommend planning the course of the interview in advance and then reflecting upon it.

With proper preparation, you become calmer, and this makes it easier for you to gain the trust of the interviewee.

A typical needfinding conversation might look like this:

1. Introduction

First, we introduce ourselves and explain the reason for the request as well as the course of the interview. In so doing, we emphasize that there is no "true" or "false" and ask whether we are allowed to document the interview (e.g., video, photos, or audio recording). The main point is to create an atmosphere in which the respondent feels comfortable. Respondents must have the feeling they are appreciated and understand that their knowledge and experience are valuable to us.

2. Actual beginning

The interviewees can also introduce themselves at the beginning, so a simple reference to the problem is easily established. We commence the interview with a general and open question about the actual theme. Based on the answer, we go deeper with questions that expand and clarify the issue. What's important is that the people questioned feel comfortable and we win their trust.

3. Create reference

We try to find a recent example that the person remembers well. This way, we bring the person closer to the topic and the problems. It might happen that not all the problems or critical experiences are expressed in this example or on the same day. Continue to build trust, assure the interviewees that their answers are important, good, and helpful to us. If the desired depth is not reached yet, we are patient and ask for more experiences and stories.

4. Grand tour

Deepen other critical topics and search for contradictions. Get to the bottom of details if possible. This can refer to both tangible and emotional facts. We have reached our goal when things that were hidden come to light. If the interviewees trust us, they can open up and share exciting stories and needs with us that would have remained hidden in a normal interview.

5. Reflection

We pause for a moment and then come to the end of the interview. We express our gratitude for the important findings and summarize the main points from our point of view. Often, the person interviewed adds important things, points out inconsistencies, and emphasizes important items. At this point, we can ask the "why" question and dig deeper, if necessary. In this phase, we are free to switch to a more general level in order to discuss explanations or theories on the matter under discussion.

6. Wrap-up

Don't turn off the recording device yet! Often, the most intriguing things occur at the very end, so we should give the end enough space and time. We thank the interviewee again for the conversation, the time spent with us, and the insights we gained. We give the interviewee the opportunity to ask us questions. After the interview, we reflect on it by summarizing the most important findings, both in terms of content and approach.

Most people are uncomfortable asking open-ended questions. The situation happens every day, and most of us are familiar with it:

Priya waits for tram 5 in front of the Pension Management Institute in Zurich. The waiting time amounts to 9 minutes. An elderly lady is standing next to Priya, also waiting for tram 5 and looking rather bored. At that moment, Priya can think of a thousand reasons why needfinding doesn't bring any added value anyway and that she will probably find another elderly lady at the tram stop tomorrow. Most of us feel the same as Priya during this phase. Nearly everybody feels uncomfortable and even a little embarrassed to approach strangers. Priya is definitely not alone in this respect. But what can really happen to her? She is merely interested in the lives of others so as to enrich her idea with knowledge and deepen her insights.

Priya works up the courage to start a conversation, but how should she begin and how would she place her questions?

Of course, Priya's primary goal should be to ask her questions in such a way that the elderly lady tells her something about herself and her exercise habits in everyday life. We have found it rather useful to create a question map in advance. Now you might justifiably ask, why not use a questionnaire like the colleagues from strategy consulting do? A questionnaire has a linear structure. We begin with the first question at the top and work our way down. In a conversation, though, we do not think and answer in a linear way but ad hoc. The map helps us visualize topical islands that provide orientation in the interview.

On Priya's theme map, one question is what motivates the elderly lady to do sports. In addition: What types of exercising inspire the lady? What does it take for her to be happy?

Now it is important for Priya to listen carefully when the lady begins to talk about her life. During the conversation, Priya should write down important information. At the same time, making notes expresses a certain appreciation of the elderly lady—an indirect compliment that she enjoys for sure.

Priya writes down the lady's statements in her exact words, such as, "I like exercising in the morning because it stimulates me mentally." If she only writes down keywords, Priya will have to make up the statements later on or invent context. Priya can then compare the statements provided by different respondents in the synthesis and recognize similarities as well as differences. She can also integrate the sentences perfectly into her persona, lending it an authentic voice this way.

After each interview and each observation, we should ask ourselves some key questions:

Where did the person reveal the biggest problems?
What is the need behind the problem?
What innovation would make everyday life easier for this person?

This is also referred to as situation-inspired ideation. We outline the ideas and thoughts that emerge directly during the needfinding. Priya could also write down supplementary questions she comes up with over the course of time when she is in different situations (e.g., whether seniors living in the country exercise more often). This way, Priya enriches her question map and extends her question horizon.

Nutrition
- Changes
- Rhythm

Seniors

Life in old age
- What changes?
- What is difficult?

Sports in old age
- Motivation

Being happy
- Significance

Tracing behaviors

"Why do you smile when you say that?"

"How did it happen that . . .?" / "Who has taught you that?"

"How do you know how it works?"

"What works?" / "What doesn't work?"

Gaining clarity

"What exactly do you mean by . . .?"

"How would you describe it in your own words?"

Exploring actively

"You say this is difficult. What exactly was/is difficult about it?"

"A difficult task. Why exactly is it difficult for you?"

Asking about the sequence (day/week/period of life)

"What is your first memory of . . .?"

"What happened before/afterward?"

"How did you do it before?"

"When was the first/last time that you . . .?"

Asking for examples

"What was the last app you downloaded?"

"With whom did you discuss it?"

Exploring exceptions

"When didn't it work, then?"

"Did you have problems with . . . before?"

Understanding connections and relations

"How do you communicate with . . .?"

"From whom did you hear that?"

"Who helped you with it?"

"How did you hear of it?"

Informing outsiders

"If you had to explain it to an exchange student, what would you say?"

"How would you explain this to your grandparents?"

"How would you describe it to a small child?"

Comparing processes

"What is the difference between your home and that of your friend?"

"What is the difference when you do this on the road instead of at home?"

Imagining the future

"How do you think you'll do it in 2030?"

(What if it were like that already today?)

The observation and questioning of lead users (users or customers leading the trend) can help to identify future customer needs. In addition, lead users can be drawn upon as another source for understanding customer needs, and their experience can be integrated in the empathy mode of design thinking.

The term "lead user" was coined by Eric von Hippel. According to the definition, lead users are users who have the needs and requirements earlier than the mass market and hope for a particularly high benefit and competitive edge from the possible satisfaction of the need or solution of the problem. Lead users have developed many major innovations themselves. These include the mountain bike, the hyperlink structure of the World Wide Web, and GEOX shoes. Lead users have a strong drive to solve a certain problem they have. This state drives them to innovations, which they often actually realize in the form of interim solutions or prototypes.

We propose an easy to follow, three-step approach to involve lead users:

Step 1: We identify needs and trends
- Scanning of secondary sources (future researchers, trend reports, trend scouting, etc.) for early trends, research directions, market experts and technology experts
- Preliminary determination of important early trends and future needs in early phases

Step 2: We search for lead users and lead experts
- Search for lead users and lead experts in the target market
- Identify analogous markets by abstracting your own questions and topics and transferring them

Step 3: We develop solution concepts
- As a last phase, rudimentary solution ideas identified thus far are finally developed into strong innovation concepts in a large workshop together with lead users, lead experts, internal marketing, and technicians.
- In the framework of co-creation, it is useful to involve the lead users heavily in the development and prototype process.

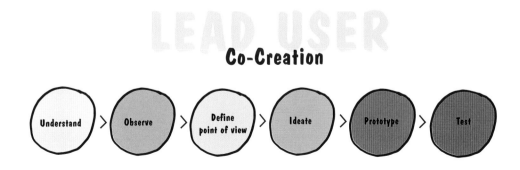

Co-Creation

Understand > Observe > Define point of view > Ideate > Prototype > Test

Lilly has read the book *Crossing the Chasm* (Geoffrey Moore). When selecting lead users and adapting the solution to the needs of the lead users, she is aware that there is probably a gap between the needs of the lead users, early adopters, and the early majority. This is why she always attempts to recognize the needs of "normal" customers in her workshops. With regard to the ultimate solution, it is then important not to forget the needs of the early majority.

Peter has been on projects that focused far too much on the needs of lead users—in the end, a product emerged that was given the sobriquet of "white elephant." Such projects have high risk and low likelihood of implementation and are hard to stop. Unfortunately, in some cases, there are only a few customers for a solution that was deemed quite interesting by many lead users.

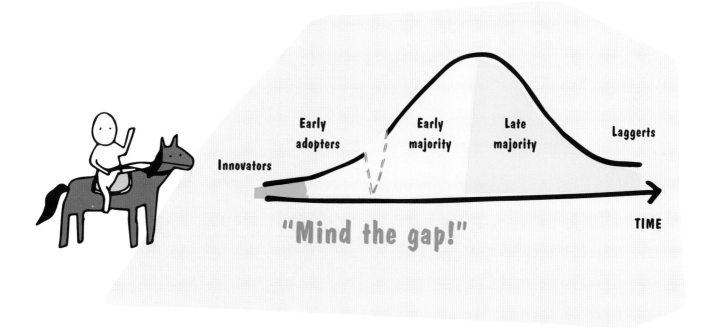

For us to be able to take a peek behind the scenes of our users, we must be able to build up a deep-rooted empathy with them. Various methods and tools that go beyond pure observation can help to achieve this. At this point, it is important to emphasize again that we can recognize the real needs of our users only when we go to work with the right attitude. We summarize:

Dig deeper

It is our goal to take a look behind the scenes. We dig deeper when searching for real needs!

We listen attentively to personal stories and experiences of the interviewees.

We set aside our own experiences. It is best to forget our own problems and wishes!

Own problems

We search for the workarounds, remedies, and quick fixes of our users.

We differentiate between needs and solutions. If we already have a solution in mind, we find the problem and the need for the solution!

Alternative solutions

One solution

Need

We recognize contradictions between what interviewees say and what they actually do.

I cannot walk on stilts!

Problem

As we have seen, the WH questions help in the divergent phase to gain a basic overview and in-depth insights. WH questions help to get better information, thus comprehend the problem or situation better.

What	Who	Why	Where	When	How
What is the problem?	Who is involved?	Why is the problem important?	Where does the problem occur?	When did the problem begin?	How could this problem be an opportunity?
What would we like to know?	Who is affected by the situation?	Why does it occur?	Where was it already resolved before?	When do people want to see results?	How could it be solved?
What are the assumptions that are scrutinized?	Who decides?	Why was it not yet solved?	Where did similar situations exist?	When can the project be started?	What has already been tried to resolve the problem?

Where? Who? How? When? Why? What?

Especially in the first few design thinking phases, the WH questions are of vital importance.

They help us make concrete observations in a specific situation and thus discover more emotions and motives. In addition, the WH questions help analyze and scrutinize information already gathered.

1. Create a set of WH questions.
2. Make a list of possible sub-questions.
3. Try to answer all WH questions.
4. If a WH question does not make sense in the context of the problem, skip it.
5. If the WH questions were used with the user in the context of a problem interview, try to dig deeper by probing and repeating questions.
6. Try to find more than one answer to every question. Conflicting answers can be of particular interest and should be amplified more deeply together with the user.
7. Evaluate the answers only at the end and filter the statements according to their relevance to the solution.

HOW MIGHT WE...
reflect on our own behavior and assumptions?

The task now is to reflect on what we have heard and seen as well as our own behavior. This transition helps to improve the process continuously.

The reflection proceeds along three steps:

First step: Reflect upon the user and the need. What have we learned in relation to the project?
We ask ourselves these project-related questions:

- How do people think and act in everyday life?
- What is done differently than we imagined?
- What surprises us ("Eureka!" moments)?
- Is there a need that is worth being solved?

Second step: Is our solution the right one?
In a second step, we check whether our solution feels right. Is it really true that we don't have to change anything for our idea to work in everyday life? What would we change so that our innovation is used in everyday life?
Priya, for example, quickly realizes that her view of the solution has expanded by asking questions and reflecting upon things.
Now Priya's arguments are no longer based on her assumptions but on the things she has heard and seen, as well as on knowledge she has collected. She has a solid idea of how it really feels to be at an advanced age and wishing for a healthy life.

After having expanded the perspective on a topic, we can now go back to ideation. We iterate our original solution based on interactions with the potential users. An iteration means to improve something in an existing idea or to build a completely new prototype.

Third step: Were our approach and the kind of questions right?
In a final step, we check whether our approach was right. Did the way we posed our questions come across well? Was our documentation of any use later? This way, we see what was good, where we should improve things, and what we still should try out.

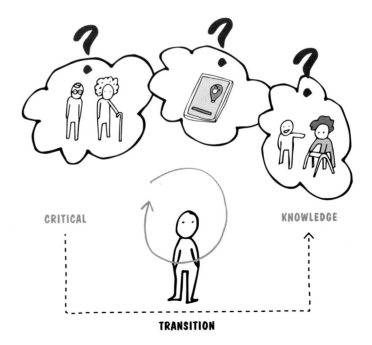

KEY LEARNINGS
Recognizing user needs

- Find the persona in real life and interview them.
- Forget all initial assumptions on ideas for a product or service and focus on the user behavior.
- Observe and listen carefully in conversation with a potential user.
- Document the observations exactly so you will be able to correct assumptions made.
- Walk in the user's footsteps; accompany him for a day in his daily life.
- Identify extreme users as well; for example, there are seniors who do strenuous sports even at a very old age.
- Review your experience with other users iteratively and never stop being curious about their real needs.
- Plan and prepare the needfinding interview diligently. Create a question map for an open interview.
- Ask many WH questions and pay attention to contradictions in the answers. Use the 6 WH question method.
- Pay attention to the end; it might still yield important insights.
- Use lead users to recognize future needs at an earlier stage—selection of the right lead users is crucial here.
- Take a look behind the scenes; dig deeper and combine various methods, such as participatory observation and extreme user or expert discussions.

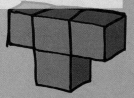

1.5 How to build empathy with the user

In the needfinding revolving around the theme of "health for seniors," Priya has realized how important it is to develop empathy for a target group. *Empathy* is the ability and willingness to recognize and understand the thoughts, emotions, motives, and personality traits of another person. By definition, design thinking is an empathetic, optimistic, and creative way of working to shape the future. When we look at any number of offers for seniors on the market, we can see that frequently neither empathy for seniors nor an optimistic basic attitude exists. Retirees do not want to be referred to as "generation 65+," "best agers," or as a target group in the "silver markets." Neither do they want to book a trip for seniors on the Internet or be invited to "exercises for seniors." Retirees are not interested in illnesses. They want to stay healthy and mobile. In most cases, they feel up to fifteen years younger than they actually are. If you don't want to make the same mistakes and in the end deliver a brilliant performance with a major flop on the market, empathy with the users is elementary.

How can we build empathy with the potential user (by way of the example of "seniors")?

Priya already has an idea for it. She has developed a prototype smartphone for seniors. It has a simple interface to a blood pressure measurement device. The prototype, named the "ImedHeinz," is a little clunky, due to the large keys and an analog interface to the blood pressure measurement device. The enclosure for the smartphone is reminiscent of the smart pocket calculators from the 1980s.

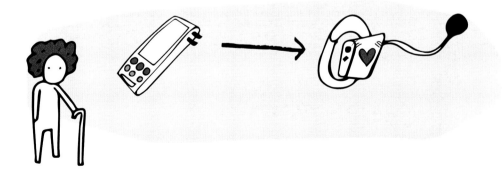

Priya wants to test the prototype in an environment of seniors and pays a visit to the "Shady Pine Tree" retirement home. In the dining room, she meets Anna: 70 years old, fit mentally, using a wheelchair due to a stroke, which has prompted her to move from her townhouse to the retirement home. Priya confronts Anna with the prototype of her ImedHeinz smartphone. Anna's response is a horrified look. To excite a little enthusiasm in Anna, Priya shows her, quite euphorically, how quickly the data from the blood pressure measuring device can be transmitted to the Heinz. Anna does not show any enthusiasm whatsoever.

This brings Priya down to earth somewhat; she leans back in her chair, and her gaze wanders to the other seniors in the dining room. Richard, sitting at the back of the room, is playing chess on his tablet; Elizabeth is exchanging WhatsApp messages with her grandson in New York City on an iPhone. Anna takes Priya's hand and says she is a great iPhone fan, too, and that she is looking forward to her new, gilded iPhone that would match her jewelry so well.

Priya has learned a lot this afternoon. The basic prerequisite for empathetic needfinding is the immediate proximity to the customers (seniors) as well as the readiness to engage with your interlocutor and to try to experience the world through another person's eyes. It takes courage and strength to step back from known standards and views of the world—but without it, needfinding and the empathy with a potential user it requires can hardly take place.

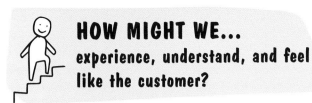

HOW MIGHT WE...
experience, understand, and feel like the customer?

UNDERSTAND THE LANGUAGE OF THE CUSTOMER

Misunderstandings are often based on everyday problems. Depending on many aspects such as family background, lifestyle, values, and context, people think and act in the most varied ways. Perceiving these nuances from the start gives us insights into the lives of our users that constitute the cornerstones of successful innovations.

How can we better understand the language of the customer?

We listen actively and ask about words that can be understood in different ways. For example, what do we mean when we speak of "resources"? The word might refer to time, material, or even people. We can only know what our interlocutor means if we have him explain it to us.

If we observe some conspicuous theatrics, the time has come to dig deeper. If our interlocutor talks about an "incredibly exciting" situation, for instance, but rolls his eyes at the same time—what does it mean? We should always ask, "We could see how you rolled your eyes. What did you mean?" From our experience, we know that a better understanding of the customer's language and personality also results in a better understanding of his need.

2 EXPERIENCE THE WORLD OF THE CUSTOMER

Instead of endlessly speculating on users' everyday situations, it is far more instructive for us to experience them ourselves. This way, critical facts crystallize that can serve as starting points for innovations. But be careful: We experience only a fraction of them.

How can we experience the world of the customer?

To recognize critical needs, we adopt the perspective of the user. This requires empathy. Our own thought patterns and principles can inhibit us, because we innovate not for ourselves but for the user. In an authentic environment and with empathy, the world can be experienced through the eyes of those people who will use our product or service day after day.

3 POLLO EFFECT/ HAVING AN OPEN MIND

None of us wants to ask naive questions, which cause other people to roll their eyes. We might speak of the "pollo" effect in this context (*pollo* is Spanish for chicken). None of us wants to volunteer to be the chicken; however, this is often necessary to immerse oneself in the world of the user. Dare to ask questions because there are no naive questions.

How do we manage to keep an open mind?

It is important to ask questions with an open mind and put our personal experiences and values to the side as much as possible.

The best approach is to imagine that we are aliens in this world, apprentices from a strange galaxy. We have never been here before and we don't know how people live here. Our lives are completely different, and everything we hear is new and inexplicable to us. As friendly extraterrestrial beings, we try to ask users questions in a nonbiased way. This way, we discover their world and their behavior as if it were completely new. As one would expect, the statements of the respondents are more differentiated this way because the role of apprentice is not threatening to the interviewee. On the contrary, in our experience, curiosity motivates potential users to tell us even more.

♛ EXPERT TIP
Mindfulness as the basis of empathy and innovation success

Mindfulness is a basic capability of our brain, but our ambition to multitask makes us suppress this condition too often in our everyday lives. If we focus our attention purposefully, we are able to be more precise with our perception and be more present. Mindfulness may be directed both inward and outward. Empathy is best built up when we focus on the current moment, involve all our senses, and perceive the situation in an unprejudiced way. Mindfulness is the basis for enhanced cognitive skills and is a vital aspect of the design thinking mindset because it stimulates and promotes our creativity while cultivating our empathy and emotional intelligence. The things we experience with great enthusiasm and attention, we internalize better.

Why is it so hard for us to be empathetic?

It seems that society as a whole is getting increasingly incapable of empathy. This is probably because in our achievement-oriented society, we are exposed to a constant pressure to optimize ourselves. If our interlocutor seems strange and foreign to us, or if we have already sorted him into a certain category of people by our interpretations and conclusions, we will have a hard time empathizing with him. Other influencing factors also affect our empathetic behavior: a stressful everyday life; pressure to succeed; hectic situations; being wiped out; and emotional states such as anger, rage, and fear.

Behind these emotions, you usually find unfilled needs, unconscious beliefs, prejudices, and evaluation patterns. All this reduces our willingness to put ourselves in others' shoes and ultimately blocks our empathy.

Full mind or mindful?

Mindfulness as the key to empathy:

1. Dare to change the perspective: Observe the world from the other side.
2. Give full attention to a topic: Be mindful, present, and precise.
3. Listen attentively and actively: Use looks, gestures, and facial expressions to be present.
4. Reflect on your own behavior: How do I come across to other people?
5. Read signals: What do the facial expressions, gestures, and voice of your interlocutor tell you?
6. Question your readiness for empathy: Is my opinion without prejudice?
7. Ask open questions: How might the future look?
8. Explore feelings and needs: How do you feel today?
9. Express your own feelings and needs: "I wish . . ."
10. Act with empathy: How can I help?

What types of empathy are there?

Empathy is important. Like everything else in life, we can break down this basic human trait into various stages. The terms "emotional intelligence" and "emotional empathy" are often used in this context. This intelligence of emotions is becoming more and more important, whether in product design, employee management, or in human relationships. It is the ability to perceive other people's emotional sensitivities and respond to them adequately. While cognitive empathy merely allows us to recognize, in a first step, what another person feels, emotional empathy allows us to feel what the other person is feeling. The strongest form is when we suffer with the other person mentally and physically.

How can we train empathy and mindfulness in everyday life with the help of a talking stick?

The talking stick, adopted from Northwest American indigenous cultures, is a tool for empathy and attention. In a meeting, the talking stick is given to one person. This person explains his standpoint and keeps the talking stick until he feels the other participants in the meeting have understood him. The other participants listen or ask comprehension questions; otherwise they remain silent.

What is the benefit of the talking stick?

The talking stick promotes empathy because the other people in the room listen until they put themselves in the position of the speaker and are able to give the individual speakers the feeling that they have been understood. In everyday working life, this can yield vital benefits:

- Members improve their ability to listen.
- Being understood boosts the willingness to compromise.
- Members' ability to change perspective is nurtured and promoted.
- Everybody gets a chance to speak and is allowed to finish what he has to say.
- Only one person speaks at a time, which has a positive impact on acoustic comprehension.

Initially, this technique will demand more time in a meeting. As soon as the talking stick has been established, you will feel an increase in empathy.

In the digital environment, empathy has become a pivotal element for linking context with emotions. Usually, emotional states such as love, laughter, joy, surprise, sorrow, and anger are used. Facebook is a good example. Alongside the well-known "Like," five other emoticons can be used. With the heart, users can express their love for things or persons; a smiley face is intended for funny contributions, and the wide-eyed emoticon stands for surprise. The two remaining emoticons stand for anger or sadness, either looking sternly at the world under a flushed forehead or crying.

Empathy buttons

With such emoticons, comprehensive data analyses on content can be carried out. So far, Like buttons have been integrated on product and service sites many, many times throughout the world. This data was only binary but has already yielded deep insights into user behavior and preferences. With emotional emoticons, services, products, and their users can be analyzed on an even more granular level. In the area of UX design, emoticons have another advantage: In many areas, users find it too burdensome to comment with a written text on a certain content, **so a lot of content fades on the net** without being rated. Emoticons can be easily entered on smart watches and mobile devices, which increases the comment rate considerably. Simplicity is becoming more and more pivotal!

Why does simplicity convince users?

Until 2012, dating sites consisted of nothing more than screens filled with profile pictures in grid design. Although targeted at the millennials (generation Y), this type of depiction did not meet their individual demands for flexibility, efficiency, and autonomy. Swipe right and swipe left changed our user experience fundamentally! The developers of Tinder wondered at the outset whether it made any sense at all to provide another service in this fiercely competitive segment. Their idea emerged from complaints received from the users of other dating sites. These users' needs constituted the basis both for the design process and the selection of the features implemented.

Tinder's entire concept is focused on the mobile user experience. Most functions of conventional dating platforms were purposefully omitted, and interaction options were reduced to a bare minimum.

Tinder is an example of three core properties of empathy in UX design:

1. Personal bond:

One picture per page and a simple interaction possibility ensures a personal and self-contained UX. The process is efficient, and the service can be called up anywhere. If there is an interest, it is even possible to obtain more information about the potential match with a click.

2. Motivation:

Because a match comes about only if both searchers signal their interest, users quickly undergo moments of happiness. Such moments have a strongly motivating impact and ensure long-term customer loyalty (hook effect; see Chapter 1.1).

3. Trust:

The chat function strengthens the trust of users in the app. The virtual date turns out to be not only surface sham—there is actually a possibility for users to arrange to meet for a quick coffee around the corner.

Good UX designers need to know how user groups come in contact with specific technologies and why they interact with them. For responding appropriately to users' emotions, however, empathy is more important. Before beginning with the design of a product, we as designers must be in contact with the users in social networks or in the real world in order to get an authentic picture of their behavior and needs.

How can the love and passion for a product win over the users?

The credo of Lingscars.com is probably the opposite of simplicity. Ling Valentine lives for cars, for affordable leasing, and especially for her customers. Her start-up came into being in 2000 when she realized she was far more capable of operating the business than her husband. Her recipe for success is to let an emotional relationship emerge, which is of fundamental importance if you want to sell cars successfully. Most major leasing companies cannot offer this customer experience. To achieve that, Ling broke all design rules. Her Web site radiates powerful colors, different fonts, and unique graphics. This approach attracts hundreds of thousands of new users to the Web site every month. Lingscars.com was described by *Management Today* as "the most cluttered Web site we've ever seen." She won awards for the ugliest Web site of all time and recognition for the large number of visitors. Ling is close to her customers, whether in stunts in which she appears or through her presence on blogs and the social networks.

Ling lives from viral marketing and the "word of mouth" of her satisfied customers. She addresses her customers directly. Her Chinese military truck—equipped with a huge rocket and the ad for Ling's Cars—is a daily attraction for drivers passing it on the highway. She passes on her simple and cost-effective marketing 1:1 to her customers with attractive conditions. Her customers love her for it.

The cultural differences in the perception of design and how design thinking is used in the respective culture itself must always be taken into account.

KEY LEARNINGS
Build empathy with the user

- Build empathy by understanding the actual needs and backgrounds of potential users.
- Observe potential users without prejudice and in their actual environment.
- Act like an extraterrestrial being who has entered a new galaxy for the first time.
- Improve empathy by perceiving your own wishes, which in turn makes you more open to the needs of others.
- Listening carefully is a crucial component of empathy. Pay attention to the body language (nonverbal communication) and probe if it seems to contradict what has been said.
- Transfer emotions about contents from the digital world to reality by emoticons.
- Draw conclusions from variations of the emoticons about users, their behavior, and their emotional relationship to content, products, and services.
- Enhance user experience—even of a digital product—through the empathy you have built up.
- Make sure that all parties involved in the development phase (e.g., UX) are already dealing actively with users' wishes.
- Pay heed to the cultural context in UX because it can strongly affect how the user perceives the offer.

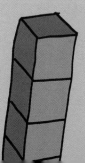

By way of introduction, we already pointed out in Chapter 1.2 that the hardest thing is to determine the point of view (in short: PoV). Therefore we would like to introduce tools and methods that make this step easier.

Peter, Lilly, and Marc are frequently faced with the challenge of having to find a solution not only for one group of people but also one that is relevant to a multitude of different users or customers. In such cases, it is decisive to assume a 360° view.

In principle, empathy with a potential user is an important integral part when preparing the ideation phase; it also reminds us of our limits. Empathy is vital not only for selecting the right community but also for the way in which we pose the right questions during this phase. The questions should prompt interviewees to put themselves in different situations and consider them from different points of view.

What might the sequence for this look like? It starts with formulating the problem; then comes the definition of the relevant points of view, which ultimately leads to the questions being answered within the set framework.

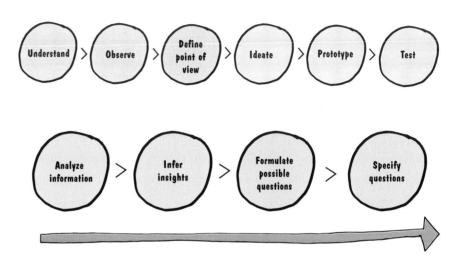

Up to now, we have concentrated on developing a product for a single group of people or users and emphasized how important empathy is. Now we want to go one step further and solve a problem for a wide range of users. The concrete procedure is described on page 81.

In our experience, the following approach is suitable for obtaining a good PoV.

A) Analyze information

- Collection, interpretation, and analysis of all information.
- Summary and consolidation of key findings into insights.

B) Infer insights

- Summarize the 10 most important insights.
- Infer the design principles or problem clusters from it.

C) Formulate possible questions

- Mark possible key themes or questions (e.g., dot voting on insights and principles).
- Choose three thematic areas and formulate the question.

D) Specify questions

- Present, discuss, and select a question.
- Refine and improve the question.

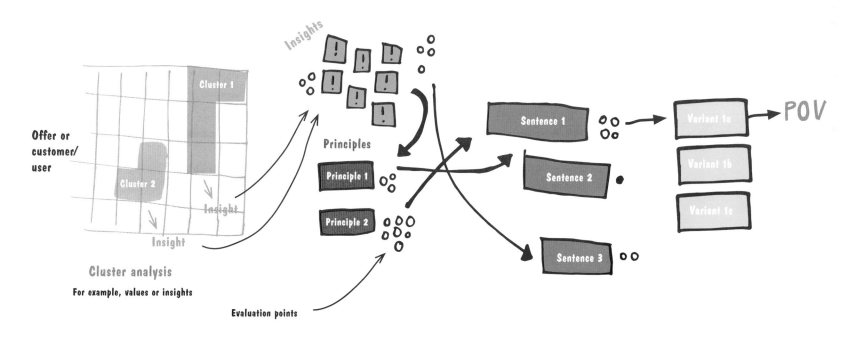

How might we solve problems for a wide range of users and address their needs?

We have had good experience with a 360° view of a question.

Let's take a need of our persona, Lilly. Lilly wants to marry Jonny. They plan their wedding party together. They do not yet know how the wedding party should look. So the problem statement is:

Lilly and Jonny don't know yet how their wedding party should look.

The question derived from this would be:

"What should Lilly and Jonny's wedding party look like?"

Based on this, the stakeholders and points of view are defined:

"Suppliers and the budget, for instance, are important when you plan a wedding."

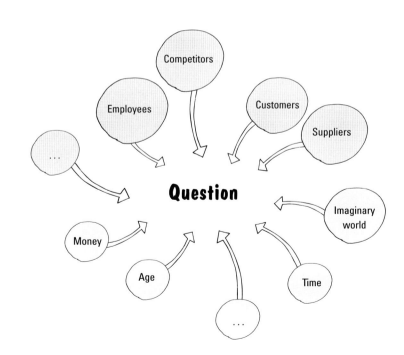

Before we search for specific ideas, the ideation phase must be well prepared. As we have learned in the previous chapters, we are dealing with the divergent phase here, meaning nothing else than that the horizon is broadened and new ideas are searched for. To allow for this, we consider the problem from as many different perspectives as possible (i.e., from a 360° view), starting with the question.

As shown by way of example in the table, the points of view range from "money" to "age." A rough differentiation is made between the stakeholders and other points of view. Of course, the number of possible points of view is infinite—hence the selection presented here is only an example. Note that if a stakeholder map exists (see Chapter 3.4, p. 258), it can also be used as a starting point.

Back to our actual problem statement: the wedding party. We want to identify the relevant points of view and reflect upon the question of which stakeholders and which other points of view might be relevant for the question of Lilly and Jonny. As shown in the table, it is useful to define the points of view and their intentions. This makes it easier to share the points of view with other participants.

Stakeholders	
Customers	- Loyal customers - Occasional customers - Noncustomers
Partners	- Suppliers - Lenders - Sponsors
Employees	- Long-standing employees - Employees with expertise - Critical employees - Apprentices
Government institutions	- Municipality - Social services - Local employment office
Residents	- Private households - Other companies or shops
Competition	- Direct competition - Indirect competition

Other points of view	
Time	- Past - Present - Future
Money	- Without money - With a lot of money
Imaginary world	- On another planet - In a fairytale - In a movie
Age	- Children - Teenagers - Adults and seniors
Culture	- In other cultures - The culture of the bride - The culture of the bridegroom
Geopolitics	- In other countries/ with other systems
Time pressure	- Without time pressure - With great time pressure

The wedding party points of view	Explanation of the point of view
The couple	Lilly and Jonny stand at the center of the question
The parents of the couple	They are very close to their children
The witnesses to the marriage	They are close friends of the couple
Children	Many families with children are invited
Seniors	Many seniors are invited
Money: With a lot of money	So dreams can be formulated
Money: Without money	So the little things that cost nothing but give a great deal won't be forgotten
On another planet: on Venus, the planet of love	For some kitsch and utopian fantasies
In another culture: in the family of the Russian Tsar	Inspiration from other cultures
At another time: in the Middle Ages	For great down-to-earthiness
Reverse point of view	To find out what the worst scenario would be

After Lilly and Jonny's point of view has been defined, a question should be formulated for every point of view. The question has the aim of enabling the potential friends, whom the couple will ask for advice, to take the said perspective and force them to answer the question from this point of view.

The questions for the ideation phase are often very broad. Lilly and Jonny won't organize a workshop in their case, but they will probably collect answers during a shared dinner with their friends or via social media/e-mail. For a "working environment," a physical workshop is advisable because creativity might suffer in a digital workshop, although feedback can be collected quickly in the latter.

To forestall respondents' expressing only those ideas the couple is willing to hear, some of the ideation should be done on an anonymous basis, such as in writing or by means of an online tool. Anonymity is not absolutely necessary for collecting great ideas—after all, it's fun to talk about things that are pleasing. But if you want to know what might bother people, anonymity is a must. To the question of "What would be the worst wedding party?" some of Lilly and Jonny's friends responded they would fear having to sit at the same table with the same people all night long. Some friends are horrified at having to wear a suit all day long. Many families would like to stay overnight locally but cannot afford an expensive hotel.

The wedding party points of view	Questions
The couple	What does the couple wish for the wedding party?
The parents of the couple	What do the parents of the couple wish for the wedding party?
The witnesses to the marriage	What do the witnesses wish for the wedding party?
Children	What do children wish for a wedding party?
Seniors	What do seniors wish for a wedding party?
Planet Venus	What would a wedding party on Venus look like?
In the family of the Russian Tsar?	What would a wedding party in the family of the Russian Tsar look like?
In the Middle Ages?	What did a wedding party in medieval times look like?
Reverse point of view	What would be the worst wedding party?

There are countless methods for adding structure to the insights: Venn diagrams, mind maps, system maps, cluster analyses, customer journeys, and so forth.

The 9-window tool is a simple method for analyzing potential application cases and customer needs. In so doing, the product or service is more closely examined in the dimensions of "system" and "time."

"System" refers to the structure of a product or service, including its entire environment. It invites you to zoom in to the product/service (subsystem) or to consider the super system (zoom out).

In the dimension of "time," we vary the temporal consideration and focus on what happened in the past or might happen in the future. This approach helps us overcome barriers and see the product or digital service from a different point of view.

With the 9-window tool, Marc can structure his business idea about the theme of "patient record" (example 1).

Example 1: IT development "Health" patient record

Traditional

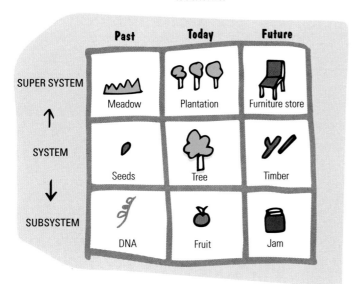

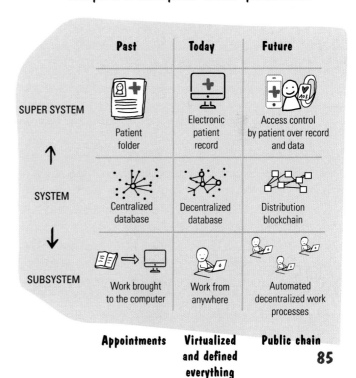

Jonny, who predicts that the banking landscape will change in the way they lend money, can transmit the effects of various blockchain evolution levels to the respective subsystems and super systems (example 2).

Frequently, the many elements are prioritized, such as by means of scoring. The elements with the highest score are pursued and one or several are chosen for the PoV.

A daisy map can be used to depict the most important elements. Its advantage is that the most important items are highlighted, so it's not always the top item that is automatically seen as the most important. All five to eight flower petals are equal, as it were.

Example 2: Blockchain finance monetary system

	Past	Today	Future
SUPER SYSTEM	¥ £ $ Currency	$ Fiat exchange	Autonomous decentralized monetary system
SYSTEM	₿ Blockchain 1.0	Blockchain 2.0	Blockchain 3.0
SUBSYSTEM	(\$) Payment	Initial Coin Offering	Digital assets

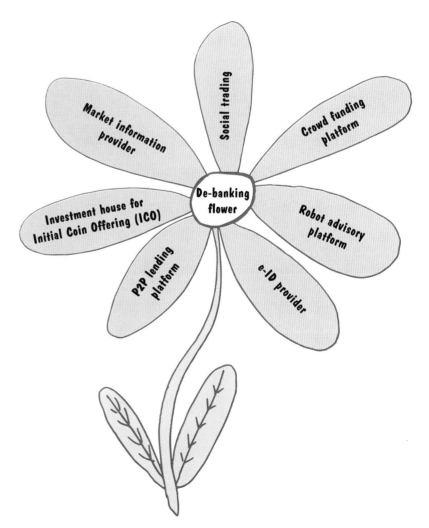

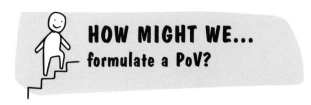

HOW MIGHT WE...
formulate a PoV?

As we could see from the example of Lilly and Jonny's wedding, the PoVs serve mainly to collect, structure, and weight all insights so as to find the relevant points. They also help us identify contradictions and determine the priorities for the next iterations. This is referred to as synthesis.

Synthesis is about finding the important needs and patterns of users, including those that were undiscovered up to now. The result of our synthesis is one condensed sentence, the PoV, which determines the question for the coming ideation phase. We will return to the topic of the synthesis in this expert tip because it poses a great challenge for many design thinking teams.

Every PoV sentence is a starting point that will be adapted in the next iteration lap.

What do we focus on in the PoV phase?

- We recognize patterns in the needs of users.
- We see opportunities where others see problems.
- We understand the needs of our customers at all levels.
- We provide clarity about assumptions and hypotheses.
- We immerse ourselves in systems and make them tangible.
- We consolidate information and interpret it.
- We understand findings and emphasize the most important insights.
- We create the starting point and focus on the PoV for the next ideation.

We recommend formulating the PoV in a catchy sentence. Use various formulations. Try and test which variant is best for you, the team, and the situation.

Approach	PoV sentence/fill-in-the-blank text
How might we	How might we . . . e.g., in the form: How might we help [the user, customer] to achieve [a certain goal]? Or: How many ways are there to achieve [a certain target] for [the user]? Example: How might we help patients keep their health records safe and share them with a doctor at a given time?
Stanford PoV	[User] needs to [need] because [surprising insight]. Or: [Who] wants [what] for [need fulfillment] because [motivation] . . . Example: The patient must have the data sovereignty for his health data because he wants to avoid abuse.
Agile methods User stories	As a [role/persona] ("who") I would like to [action, destination, wish] ("what"), in order to achieve [benefit] ("why").

KEY LEARNINGS
Find the right focus

- Try to find the point of view with a 360° view.
- Look at the situation from different points of view and define the focus for the next iteration.
- Use the 9-window tool to explore what happens before and after the use of the product as well as what is happening in the system.
- Present the needs not in the form of a list but as a daisy map.
- Change perspective, e.g., "time" (before, after), "money" (with, without), etc.
- Use a fill-in-the-blank text in different variants, which can be changed according to project, maturity, preferences.
- Start a project with pretty simple WH questions: "How might we. . .?" or "How many different ways are there for. . .?"
- Always develop various PoV questions and choose the most suitable one from among them.

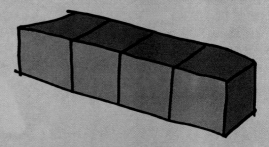

1.7 How to generate ideas

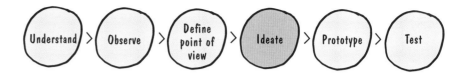

Without ideas, no new products! The importance of finding good ideas at the right time is enormous, putting both participants and workshop facilitators, whose job is to tease out ideas from attendees, under pressure.

We know from studies that groundbreaking ideas don't always emerge during a brainstorming session; sometimes, the creative spark leaps while you have a shower or scribble something on a napkin. This is why creative companies give their employees more and more leeway to allow for this type of intrinsic inspiration to happen; for instance, in the form of workdays on which employees are allowed to do whatever they want. The only condition is for them to report back what they have done.

However, often the milestones already have been set, and as product developers or engineers we are in no position to explain to the boss that we want to spend the next four hours in the shower because the chances of hitting on great ideas are better there. So we need methods and tools of structured ideation.

Peter in particular is under a lot of pressure due to deadlines. He must deliver up creative results and, consequently, get his team to cough up creative outputs and put people in the right mood at the touch of a button. Lilly knows from experience that some factors must be met for this touch of the button to be effective. The following credo actually seems too banal and childish to her: "A good mood is the #1 pre-requisite." Despite this, she's convinced that the potential of shared ideation can only unfold when a casual, relaxed atmosphere prevails. Only then can attendees engage in a search for ideas on a broad basis. The switch to a different or new environment alone can change the mood. If the meeting takes place week after week in the same conference room that is associated with some boring statistics, it is not conducive to a good atmosphere. So why not move the workshop to another room, outside, or even to the closest bar?

EXPERT TIP
Rules for a good brainstorming session

Before beginning with any brainstorming, people must laugh at least once. A warmup that makes participants smile helps. From our experience, it's best when they smile at one another.

Thinking in hierarchical structures is a hindrance to free and unfettered ideation. An apprentice does not want to make a peculiar impression on his boss when expressing a fanciful idea.

For this reason, we are encouraged to point out that the assistant and the accountant, the CEO and the marketing officer of the company, can all make an important contribution in the process of ideation. If participants do not know one another, so much the better! Not having general introductions before the brainstorming session, which would include announcing who has which role, has proven useful indeed. A nonbiased dialog is of great value.

When we feel there is a steep hierarchy in our company, we can try out the reverse approach: We form a team only from trainees, for instance, so they will have the opportunity to raise their profile and show others their creative potential. In the next workshop, the groups will then almost certainly mix at their own initiative.

The beauty of brainstorming is everybody is given the opportunity to come up with good ideas, no matter which function or role he or she has.

What are the rules we comply with in a good brainstorming session?

Brainstorming rules are numerous. Our top three are:

Creative confidence
We express all ideas that come into our heads, no matter how silly they might appear to us. Maybe the next person can base another idea exactly on our "silly" contribution. For this to work, we need the relaxed atmosphere just described.

Quantity goes before quality
Very, very important! The point of this phase is to fill the hat with as many ideas as possible—evaluation comes later. We resist the temptation of being satisfied with the first good idea. Maybe an even better idea is only five minutes away in our brainstorming session.

No criticism of ideas
Under no circumstances are ideas allowed to be criticized during this phase. The evaluation of the ideas takes place later in a separate step.

Quite conventional ideas usually mark the beginning of a brainstorming session. Their novelty value is low.

Peter has had the experience of some of his colleagues coming to every workshop with a fixed idea of how the solution might look. During the brainstorming session, it is hard to pull them away from these fixed ideas, and they generate little that is new. For this reason, Peter always holds a first session at the beginning, which he refers to as the "brain dump." All attendees have the opportunity in this session to dump their ideas so they are open to new things.

The actual search for ideas only begins in the second step. Peter encourages the participants to break out of their usual thought patterns so they can come up with some "wild" ideas. He uses two specific tricks; here's how we implement them in our workshops:

1) When we moderate a workshop with several groups, we can shape the search for ideas as an internal contest. We stop the brainstorming session after halftime and request that the groups state the number of collected ideas.

For the individual teams, this is an incentive to catch up, so they will inevitably have to venture in the direction of "wilder" ideas if they have undermatched the creative performance of the other groups. This approach allows us to see which group is wrestling with difficulties. If one group is far behind in their number of ideas, we watch to find out exactly what inhibits the team. Usually, it turns out this group has—against instructions—begun to discuss and evaluate the ideas.

2) We have the groups present the two best and two dumbest solutions they have generated. This moment is a valuable experience for every group. First, the task will induce a few giggles, which is quite a help for creating a positive atmosphere. Second, and far more important, now a debate is launched on whether some of the ideas are actually as dumb as had been assumed at first. Every dumb idea has potential! When we know how to reverse the idea successfully into something positive, we will gain valuable perspectives with a guaranteed novelty value.

Problem reversal technique

The problem reversal technique is Lilly's favorite method when she asks students to generate ideas for something but they don't really have any desire to join in. Lilly reverses the question and asks, for example, "How would you prevent creativity on your team?" The problem reversal technique stimulates creativity and gives participants the opportunity to have fun with a topic. In a second step, every negative statement is reversed into a positive one.

We must emphasize, though, that this method is less suitable for finding new product ideas. The reversed question, "What would something have to be like?" often results in a requirements list instead of ideas. We have nonetheless had good experience with the problem reversal technique; for example, for the revision and/or improvement of service processes.

Requirements versus ideas

Lilly learns that students in the technical area in particular have great difficulties finding "real ideas." They have a hard time differentiating between requirements and ideas. In a brainstorming session for a new headset, participants wrote "ergonomic," "lightweight," and "user-friendly" on their Post-its. Those participants coming from business administration wrote down words such as "cheaper" or "cutting-edge design." At this point, Lilly interrupts and explains that these things are not actually ideas but requirements for the product. Of course, we must also be clear about the problem for which we want to generate ideas. In this case: How might we communicate in the future without cell phones? The terms "ergonomic" and "cutting edge" do not entail a solution to the problem. An idea would be that, in the future, the electronics would be implanted under the skin to communicate worldwide. A somewhat less abstract idea would be to integrate the communication in accessories and clothing, such as with Google Glass.

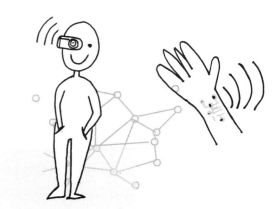

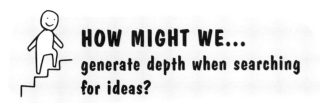

HOW MIGHT WE...
generate depth when searching for ideas?

Depth of ideas

To explain the levels of the depth of ideas and the term "requirements" better, we use the following model: We imagine we are standing in front of a ditch and want to get to the other side.

1. What is the problem? (level 1)

A ditch cuts off this side from the other side.

So our problem is that we must get to the opposite side somehow. We start brainstorming with the question: How can we get to the other side? "Safely," "in one piece," "dry," and so on, are not ideas but the requirements for the solution. They don't help us in this situation.

2. The brainstorming question (level 2)

The formulation of the brainstorming question is crucial and largely determines how many ideas can be generated or how greatly the possible solution space is expanded. Depending on the question, we restrict and channel the solution space or else expand it. The following formulations illustrate this: "What could we lay across the ditch to get to the other side?" versus "How can one overcome a physical barrier such as a ditch?"

3. Possible solution ideas (level 3)

We might "fly," "build a bridge," "beam ourselves," or "fill the ditch with so much material that we can walk across it."

4. Idea variants (level 4)

Any number of variants can evolve for each of these ideas. In a second brainstorming session, the question might be: How many ways are there to fly? "With an airplane," "with a flying bicycle," "with bird wings," "like in the Red Bull ad," "by pole vaulting," and the like.

If one group finds it hard to get away from requirements, it is advisable to have them build rudimentary models of their "ideas." This will make it mandatory for requirements to be implemented as an idea.

Tips for depth of ideas:

- We formulate the brainstorming question so it matches the solution space we want to open up.
- We can still adapt the brainstorming question during a workshop.
- The solutions from level 3 can be consolidated in a morphological box; more variants of partial solutions are conceivable.
- If a group has a hard time advancing to level 3, the instruction to translate the ideas into a physical prototype is often quite helpful. It forces the participants to become more specific. Implementing a physical model in a "user-friendly" way will help them engage in level 3.

"Prototyping"—building an idea as a physical model—is another creativity technique. The diversity of the provided material determines whether more ideas will emerge or not. The more odds and ends are available, the better it is. A balloon that's been discovered summons up the idea that something could be flexible and stretchable; a piece of cord reminds a participant that the thing might be portable.

The rubber dog in the prototyping material box:
More or less by accident, Lilly threw a rubber toy dog into the prototyping box. When the participants in the brainstorming session were tasked to translate their ideas into physical models, one of them found the toy and was highly amused by it. He started to spin ideas: "The dog could do this and that in the machine," whereupon his team members joined in and came up with more ideas. The team had a lot of fun with the dog, which enabled them to break out of their habitual thought patterns and reflect upon things they had thought little about up to now. Until the very end, the dog contributed materially to the successful outcome. Since then, it has been an integral part of Lilly's prototyping box that she brings along to the workshops.

SCAMPER is a further development of the well-known Osborn checklist. Alex Osborn, a brainstorming expert, developed in collaboration with Sidney Parnes one of the first approaches to the creative problem-solving process. For ideation, the Scamper method uses—along with brainstorming—a list of questions that should provide food for thought toward solving the problem. In our experience, it is important initially to see an example and then go through the detailed questions. SCAMPER is an acronym and stands for the terms:

SCAMPER = **S**ubstitute, **C**ombine, **A**dapt, **M**odify, **P**ut to other uses, **E**liminate, **R**earrange

SCAMPER is useful when we would like to stimulate creativity and find even more ideas. Basically, SCAMPER can be used for nearly anything: for products, processes, systems, solutions, services, business models, or ecosystems. In the event that individual questions or elements are not quite suitable or obvious, that doesn't matter for the application. We simply leave these questions out.

Substitute
Autonomous vehicle
Artificial intelligence instead of human intelligence

Combine
Vehicle + scooter
Steering of a scooter and space of a car

Rearrange
Folding vehicle

Example car

Adapt
Vehicle + bird wings

Modify
Wheelchair ramp from the rear

Eliminate
360° turn instead of reverse gear

Put to other uses
Vehicle exterior as a solar panel

Substitute
What can be substituted?
What can be used in its place?
Who can be involved instead?
Which process could be used instead?
What other material could be used instead?

Combine
What can be combined?
What can be mixed?
How might certain parts be connected?
Which purposes could be combined?

Adapt
What other ideas are suggested by it?
Is there anything that is similar and can be applied to the existing problem?
Have there been similar situations in the past?

Modify
What modification could be introduced?
Can the meaning be changed?
How might the color or shape be changed?
What can be increased?
What can be reduced?
What could be modernized?
Can it be enlarged?
Can it be downsized?

Put to other uses
For what other purposes could it be used in its present state?
For what purpose could it be used if it were modified?

Eliminate
What could be eliminated?
What are the things it would still work without?

Rearrange
What other patterns would also work?
What modifications could be introduced?
What could be replaced?
What could be rearranged?

KEY LEARNINGS
Generate ideas

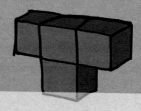

- Make sure the environment offers a good atmosphere and build up team members' creative confidence.
- Laugh a lot but never laugh at one another!
- Creativity sessions should always be in at least two parts. Provide a "brain dump" at the beginning and then stimulate creativity.
- Motivate participants to deliver a great quantity of ideas, such as via contests between teams, or by using additional sources of inspiration such as the problem reversal technique and other creativity techniques.
- Differentiate between requirements and features. Properties such as "ergonomic" or "cutting edge" are not solutions to the problem.
- Separate the ideation (the generation of ideas) from the evaluation of the ideas.
- Designate a moderator, who guides in the creativity technique, and a facilitator, who leads through the process.
- Comply with the brainstorming rules (e.g., no criticism of ideas, quantity goes before quality, etc.).
- Communicate the various ideas uniformly and objectively.
- Make use of methods such as SCAMPER, which help us to increase creativity by providing food for thought.

1.8 How to structure and select ideas

When we apply various types of brainstorming, we amass many ideas. In addition, we have consciously encouraged the teams to generate as many ideas as possible. Peter and Lilly are aware of the phenomenon: Once the team's initial reluctance has been overcome and a positive mindset has set in, ideas keep cropping up in rapid succession. Often, the screens, windows, and walls are not big enough to place all the ideas.

We recommend performing a sort of clustering initially. This can be done in different ways: Either the facilitator sets the framework, or else the teams themselves conduct a classification that seems the most suitable to them.

The agony of choice. Selecting ideas is a real challenge. For one, each one of us interprets the drawings, words, or short texts on the Post-its differently. Second, there are ideas whose basic thought goes in the same direction, and ideas that solve a completely different problem than originally intended.

The examples show that ideas can be grouped, assigned, or simply described by an umbrella term. The way is the goal, and the discussion about a meaningful classification in itself results in everybody having the same understanding of the ideas in the end. Depending on how solid the understanding and how refined the degree of detailing is, ideas can be selected directly or undergo further analysis, specification, and structuring. There are various possibilities for evaluating ideas and clusters. Having participants vote for ideas by placing adhesive dots on them is a simple way to do it. The vote is quick and democratic.

Structuring, such as with concept maps, improves the clarity of the ideas and makes it easier for the team to plan the next steps and tackle them in a focused way. Once the idea has been selected, the next step is to present it in a way appropriate for the target group. Again, there are various possibilities for this, such as creating a communication sheet of the concept ideas.

If the range of ideas is very broad and the scope of the question has been greatly expanded, the ideas can initially be grouped into overarching topics and then clustered again.

A) Matches the question

B) Exciting

C) Out of scope

Other groupings are possible:

**D) Today –
Tomorrow –
Future**

Today   Tomorrow   Future

**E) B2B –
B2C –
B2B2C**

**F) Incremental
versus radical**

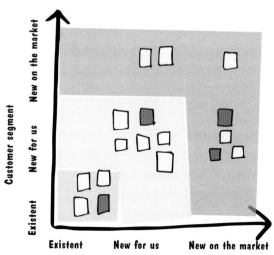

Customer segment

New on the market

New for us

Existent

Existent   New for us   New on the market

Product/service

The clustering can be used at an earlier stage to preselect the ideas, such as on the basis of the speed of dissemination and implementability, or on the basis of a Churchill matrix with the dimensions of importance and urgency.

G) Stepwise selection using the "speed of dissemination and implementability" matrix

The first step of this two-tier procedure focuses on the speed of dissemination and the speed of adaptation. Especially in a political environment—which prevails at Peter's company, for instance—it is useful to consider the decision makers and one's own influence on the potential rollout. The best ideas are characterized by rapid dissemination and fast adaptation. In a second step, the implementability and financial feasibility are investigated. This results in implementation alternatives or indications of the functional scope.

H) Selection based on the criteria of "important and urgent"

This matrix is particularly suitable when the search is for which measures to use.

We show the urgency on the x-axis and the importance on the y-axis. Then we discuss with the team which measures must be allocated to which quadrant. Once all measures have been entered, we can enter the to-do's, including the respective responsibilities, and determine milestones.

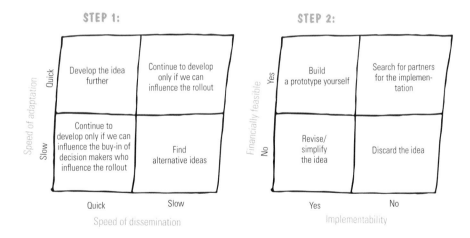

We all know the phenomenon of criteria being developed in large organizations to select potential ideas. These criteria enable a large number of teams to develop innovations in a more targeted way. Often, the criteria act as a kind of guardrail or specify certain financial goals. In general, such criteria are inhibiting, but because they do exist in reality, we should discuss them. If the criteria are not known within the framework of a well-defined strategy and vision, it is useful to ask some key questions:

- What might the vision be?
- What are the personal preferences at management level?
- What is our enterprise's culture, its values, and and sense of morals?
- Which growth areas have already been defined as part of a strategy discussion?
- What is the financial contribution an idea must yield as a minimum?
- What are the customer needs and trends on the market?

When defining criteria, it is important to be aware of reality. The potential market opportunity can be as great as it may be, but if the decision makers (e.g., top management) don't cotton to it, the idea will fail. At the least, when financial resources are allocated, we will be confronted with this situation. As illustrated, a selection via the dual matrix with the dimensions of speed of dissemination and implementability can be useful.

The values a company stands for are equally important. If it goes against the moral grain of a company to use customer data from digital channels for other business models or to sell it profitably, such ideas will have little success. Defining of these and other criteria at an early stage has at least the advantage that the waste of resources is reduced and effectiveness is boosted.

But we ought to try everything in our power to overcome such restrictions. In our experience, "submarine" projects have proven useful: projects that are initiated on the q.t. with a small number of dedicated employees and "go to the surface" only once the initial results were worked out in prototype form and have won over the decision makers.

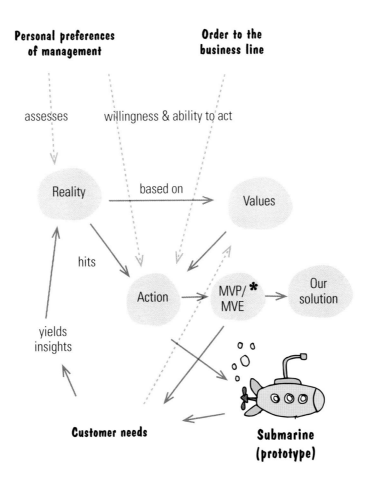

* MVP = Minimum Viable Product
  MVE = Minimum Viable Ecosystem
  (see p. 112)

Any type of structure can be depicted as a poster; for example, as a simple pro-and-con poster. The goal is to visualize the collective intelligence of a team or capture a mood. For example, we query pro and con arguments on a subject in a poster form and then have participants rate them. Lilly uses this variant to obtain quick feedback about the course from the participants at the end of her design thinking event without having to get into prolonged discussions.

For scheduling, we can draw the poster in the form of a timeline. Let's take the planning and iterative creation of *The Design Thinking Playbook* as an example. Again, we had the "courage to drop"! The party atmosphere at the end of the project was quite motivating for editors and experts. How we can use such elements otherwise and visualize them in a target-oriented way will be discussed in greater detail in Chapter 2.3.

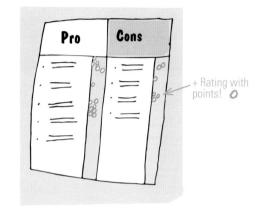

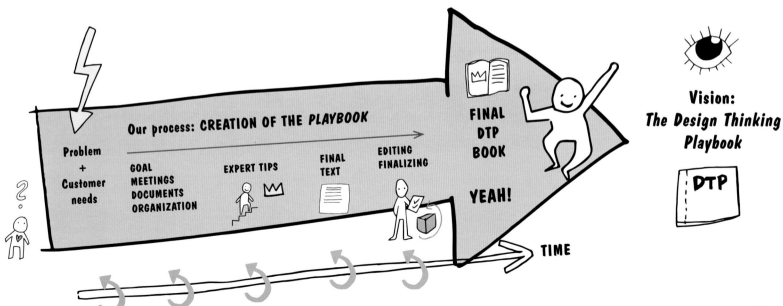

A concept map is basically nothing more than the visualization of concepts that shows us the correlations. Thus a concept map in a figurative sense is the graphic depiction of knowledge and an excellent means for us to bring order to our thoughts. The depiction of the concept map is freer than that of the better-known mind mapping.

In the mind map, the key concept is written in the center and then built up from inside to outside. Thus it looks like a tree: branches on which terms are written go off from the key concept. Hence the mind map is more a means for brainstorming. It helps to bring order to the discovered points but it doesn't show the correlations among them.

A concept map can start from several key concepts. Often there are cross-connections between the branched concepts, similar to a road network. For this reason, the creation of concept maps takes more time than that of a mind map. In our experience, at least three new creations or restructurings are usually necessary to get to a good result.

As the name suggests, a systems map is a visualization of the system. The various actors and stakeholders as well as the observed elements are sketched. Interrelationships and influences can also be depicted. In so doing, an iterative adaptation and detailing takes place. Typically, you go from the rough to the detailed (i.e., top-down). Thinking in variants is also an important element. In a systems map, material, energy, money, and information flows can be depicted. A systems map helps to understand and visualize the problem. We will address the topic of systems thinking in greater depth in Chapter 3.1. Chapter 3.3 will deal with business ecosystem design.

Other concepts, such as the giga map, will also be investigated more closely. A pragmatic description of a giga map would be "a big messy map of a big messy thing." It is also a stripped-down form of the systems map, while following the basic idea of the concept map. Giga maps can help us draw up a holistic conception of a specific task, for instance. In a final version, the giga map helps to communicate the treatise. But usually, largely owing to its complexity, it is only understood by those who created it.

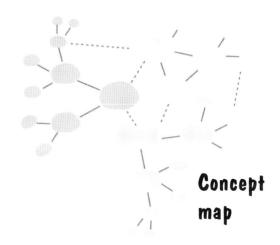

Concept map

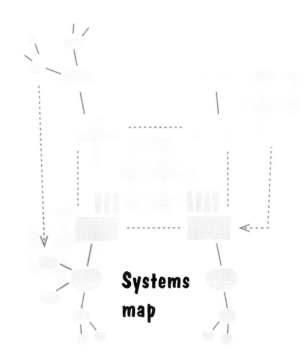

Systems map

♔ EXPERT TIP
Document and communicate
ideas with communication sheets

We work frequently with teams all over the world in major projects and organizations. The simple and clear documentation and communication of ideas is, hence, extremely important. Communication sheets on the concept ideas are a good way to achieve clarity. Ideas can be shared easily with such templates. Moreover, ideas become tangible, and possible misunderstandings are minimized.

With the compilation of communication sheets, we achieve

- the visualization of the problem and the situation,
- an improved understanding of the problem and idea,
- a better understanding of possible influences on customers and users,
- order to our thoughts,
- the recognition of approaches to the solution, and
- the documentation, summary and depiction of our knowledge.

Idea communication sheet

Name:

Slogan:

Discovered problem/need:

Sketch:

Solution:

Benefit:

Free description:

HOW MIGHT WE...
structure and select ideas?

We recommend this simple procedure for giving structure to the ideation process. First, after the ideation, bundle the ideas into clusters and structure them. Then, select the most important ideas or clusters and refine them; last, document them.
The ideas can be selected according to the needs of our persona or according to actual use.

1 Generate ideas

2 Structure the ideas

3 Select the ideas

7   14

4 Refine and document the ideas

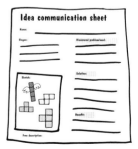

Idea communication sheet

DIVERGE

CONVERGE

KEY LEARNINGS
Structure and select ideas

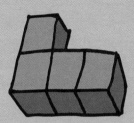

- Group the ideas and select them systematically—selection is an important step.
- Use any possibilities for structuring and visualization in order to understand and communicate problems or situations.
- Improve the team's understanding of both the problem and the solution by actively discussing the selected ideas.
- Create mind maps, concept maps, systems maps, and giga maps that enable the team to rapidly achieve a summary and presentation of the knowledge.
- Bring order to thoughts so you can better recognize approaches to solutions.
- Document concept ideas on a standard sheet to support comparability and communication.

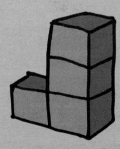

1.9 How to create a good prototype

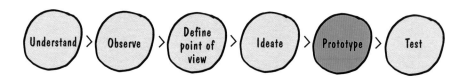

Prototyping is an important element of design thinking. It encourages us to test functions and solutions in reality, in conjunction with the desire to learn from users how to improve an offer on an ongoing basis. For this to succeed, all those involved in the process must keep an open mind so an idea can be changed or discarded. What is crucial here is the willingness to make radical changes. With a prototype, an idea is brought into a form that allows potential users to experience and evaluate it. At the outset, the prototype must be only good enough to make the relevant basic features of a future offer comprehensible to a target audience. Prototyping makes it possible to obtain quick and inexpensive targeted feedback from potential customers and users.

What is the best way to build a prototype?

Physical prototypes can be made from aluminum foil, paper, or Lego bricks; in the case of services, they can be expressed in the form of a role play. Digital prototypes can be built, designed as videos, clickable presentations, or landing pages. Of course, the various types can be combined; for example, integrating a smartphone into a cardboard box so it works as a display and thus serves as the prototype for augmented reality glasses.

First principle:

LOVE IT!   CHANGE IT!   LEAVE IT!

How does our first prototype come into being?

In general, ideas are based on many different assumptions. The task is to question these assumptions and confirm them in the real world by testing, or to discard them—that is, disprove them with a counterexample on the basis of observations and an experiment. During the prototyping process, prototypes are further developed and tested in a number of iterations until a usable offer emerges. Ideally, we begin with enough insights from trend and market research and a solid understanding of the needs and challenges of our potential customers or users.

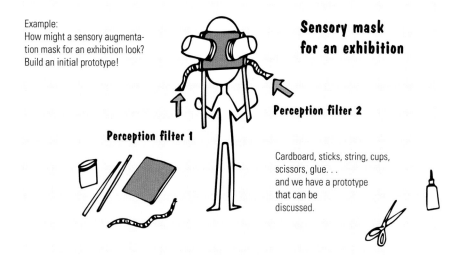

Example:
How might a sensory augmentation mask for an exhibition look? Build an initial prototype!

Sensory mask for an exhibition

Perception filter 1

Perception filter 2

Cardboard, sticks, string, cups, scissors, glue. . .
and we have a prototype that can be discussed.

An initial prototype is characterized by the fact that it can be made using the simplest available materials and as quickly as possible. The simpler, faster, and cheaper we produce a prototype, the less it will hurt us when we have to reject it. An initial prototype can be developed from cardboard, paper, plastic cups, string, tape, and other materials and then assessed.

In prototyping, a **second principle** applies:

Never fall in love with your prototype!

Over time, the degree of maturity of our prototypes will be higher, and it will be more elaborate. We must therefore schedule plenty of time for prototyping and testing. The more sophisticated our prototype is, the more accurate and meaningful the tests will be. The degree of maturity of the prototype depends on how much time and money we can invest in it. Our intention, however, must be to drive the prototype only as far as necessary in order to attain our set goals.

The test results of the prototype serve the project team as a basis for decision making in order to make the right, balanced decisions in terms of human desirability, economic feasibility, and technical implementability. Only once all these criteria have intersected are we on the right path for generating a market opportunity. We always begin with the human being and his needs.

The focus of prototyping is always on learning. As described in Chapter 1.2—in the macro cycle—prototyping is possible at any time. The added value of individual functions, a new product, or the result of a customer interaction can be tested with a prototype.

Thus we come to the **third principle** of prototyping:

It's a never-ending story: Prototyping means to iterate, iterate, and iterate still again.

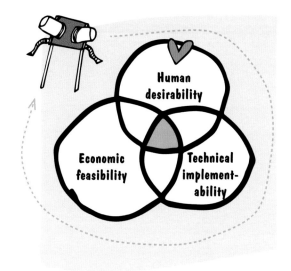

Lilly and Jonny are still dreaming of their company, which offers design thinking consultancy services. After Lilly has come up with some initial ideas for a value proposition and possible consultancy priorities, she wants to test these ideas on a potential Web site that can also be used on mobile devices. Later in the book (see Chapter 3.2), we will discuss in greater detail how a good value proposition is defined and why Lilly mainly focuses on the usability of the Web site for the time being.

Back to the prototype: Lilly outlines every page of her Web site on paper. She can't do anything wrong when building her prototype—unless she wants to get it all right the very first time, which is impossible—but she can learn a a great deal from the subsequent tests with potential users. She thinks in variants: What might be an alternative for "this or that"? She detaches herself from one variant and tries out something completely different once again—the first idea doesn't always have to be the best!

Three key questions arise when building the prototype:

- What are the basic functions for the user?
- What hasn't she taken into consideration at all yet?
- How has nobody ever done it before?

Now Lilly tests and iterates the mobile version of her Web site with potential customers until they like it and are satisfied with its navigation and the scope of its contents. She refines the prototype all the way up to the finished design, which is only programmed at the end. Generally, the following applies: The simpler the prototype of an offer can be operated, the better.

Lilly tests the mobile version of her Web site with users in an oversized version. She has made the prototype from paper.

Whatever it is we want to develop—a product, a service, an organization, a system, a space or environment, a start-up, a create-up, or a Web site—we can use different types of prototypes during the development. Our overview presents common kinds of prototypes and can encourage us as a project team to try out different things. The low, medium, or high degree of resolution (i.e., the level of detail of a prototype) helps us find out what is suitable at what point in time over the course of the development.

Degree of resolution:

low = in an early phase
medium = first approaches to a solution
high = more end solutions

Type	Description	Degree of resolution low / medium / high			Suitable for/ examples
Sketch	Paper or digital, sketched or scribbled, on a flip chart or on smaller sizes of paper such as DIN A3 or A4 (11x17 or 8.5x11 in.) or even on a Post-it.	X			Practically everything
Mock-up	Shows the overall impression of a system without its necessarily having to work.			X	Products, digital or physical
Wire frame	Early conceptual design of a system. Shows the functional aspects and the arrangement of the elements.	X			Web sites
Chart	For showing correlations. This allows you to check how ideas are linked to one another and how the experience changes over time.	X	X		Spaces, processes, structures
Paper	Building or upgrading of objects and products with paper or cardboard.	X			Products, digital or physical Furniture, accessories
Storytelling and story writing	Communication or presentation of sequences and stories.	X	X	X	Experiences
Storyboards	Shows the end-to-end customer journey of a series of images or sketches. Can also be used as a basis for a video, for storytelling, or in a funny way like a comic strip.	X	X		Experiences
Video	Recording and presentation even of complex scenarios.	X	X		Experiences
Open hardware platforms	Analog and digital interfaces for the combination with motors and sensors.		X	X	Electromechanical systems
Photo	Photo montage for the simulated depiction of a situation, using photo editing software.	X			Products, digital or physical Experiences
Physical model	Shows a two-dimensional idea in three dimensions. Can be done in the form of 3D printing, or else with other materials, e.g., Lego bricks.	X			Products, spaces and environments

Type	Description	Degree of resolution			Suitable for/ examples
		low	medium	high	
Service blueprinting	Structured description of services for the comprehensive design of the experience in the end-to-end customer journey.	X	X	X	Products, digital and physical services
Business model	Systematic depiction of business contexts and relations, e.g., with the business model canvas or lean canvas.	X	X	X	Business models
Role playing	Emotional experience of the customer with a product or service, acted by project team members.	X	X		Experiences
Bodystorming	Reproduction of specific situations, with the project team members doing the physical acting.	X			Physical experiences
Pinocchio	Rudimentary, nonfunctioning version of a product.	X			PalmPilot (personal digital assistant)
Minimum viable product (MVP)	Executable version of a system or a version, with only the most necessary functions.	X	X	X	Digital products, software
Fake door	Deliberate, fake access to a product that does not yet exist.	X	X		Zynga, Dollar Shave Club
Pretend to own	Pretending you own it (space, product, offer, etc.); actually, however, it is procured from somewhere else, has been rented or leased, before you have invested in a big way.	X	X	X	Zappos, Tesla
Relabel	Another product equipped with its own brand and packaging.	X			Products, services
Wizard of Oz (also referred to as "mechanical Turk")	User interacts with the interface of an application that does not exist. The reactions of the system are simulated by the people acting.	X	X		IBM's speech-to-text experiment
Minimum viable ecosystem (MVE)	Working collaboration on the basis of a key functionality between initial partners in the ecosystem		X	X	Blockchain applications, platform solutions (e.g., WeChat)

Degree of resolution:

low = in an early phase
medium = first approaches to a solution
high = more end solutions

HOW MIGHT WE...
better discuss our ideas and portfolio consideration with "boxing & shelfing"

During the prototyping phase it is important to make the services, products, and solutions tangible so that they can be experienced. Two methods help us to realize this: The "boxing principle" tries to use the analogy of packaging to illustrate the most important information. The "shelfing" aims to discuss a whole product portfolio and to organize the "boxes."

Boxing principle:
The basic idea behind "boxing" is to create a physical box, which can be used, for example, for the marketing of the product. Let's imagine a cereal box.

Each side of the box contains information that summarizes the benefits and characteristics of the cereal mix and the brand. The name, logo, and slogan are on the front, as well as a few points that highlight the key benefits of this brand. On the back you will find more detailed information about the ingredients and attributes of the product and some information about the company.

Core questions in boxing:

- Front: **what** is the product name, image, slogan and two to three promises about the product?
- Back: **which** details about the feature, application, and content are important?

On the remaining sides the WH questions are answered by means of text or visualizations:

- **Who** is the target customer or user?
- **Which** goals should be achieved? Which problems are solved ?
- **When** is the product available and how can we get it?
- **Where** and under what circumstances is the product used?
- **Why** should the user use the product?

The boxing principle can be used in other ways than as a product box as described. The added value of boxing is that the situation can be viewed from different perspectives. Similar to the product box, a problem box, a solution box, a project box (e.g., per project or working package), a process box (e.g., per process step) can be created.

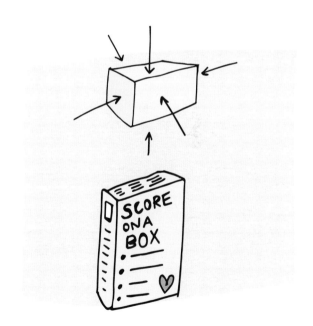

Shelfing principle:

When it comes to describing a whole portfolio, often a structure is missing for the discussion. One possibility is to sort all products, services, and solutions into three shelves: the product, service, and solution shelves.

We have had a very good experience by sorting the offerings in the respective shelves. Above the shelves, we write the categories that the customer would most likely look for. Afterward we arrange the products, services, and solutions accordingly.

The advantage of this method is that gaps in the offering portfolio, and also synergies, can be detected quickly. New ideas can be described as boxes as explained (see "boxing") and then can be sorted on the shelf. The discussion may relate to attributes such as attractiveness, novelty, strategy contribution, differentiation, etc.

The most important benefit of this technique is that it forces the team to construct their own understanding of the product in a very direct and visual manner.

This exercise provides a playful yet insightful method to pass on a deeper understanding of the product vision, while promoting discussion and collaboration between all the stakeholders.

How does our offer for the perfect kitchen look?

IKEA's portfolio in the area of kitchen can be explained well with three shelfing portfolios.

1) Solution portfolio	2) Product portfolio	3) Service portfolio
Short of space > Urban/Modern > Upscale > (Families, Singles, House owner)	Cupboards > Table tops > Electric appliances > (Families, Singles, House owner)	Planning > Delivery > Assembly > (Families, Singles, House owner)
We offer kitchens in... • Country-style • Urban for singles and families • Designs for small spaces	Our products range from... • Cupboards and tables • Products available in stone, wood, and laminate look • Refrigerators, hotplates, and ventilation equipment complement the offer	Our services include... • Planning, delivery, and installation • Carried out by qualified installation partners or for self-assembly

HOW MIGHT WE...
design a prototyping workshop?

Because we have already done a great deal of the preliminary groundwork, we assume we have developed a solid understanding of the problem statement, have verified certain assumptions, and given some thought to possible solutions. Now the focus must be transferred from the world of ideas to the real world.

Possible steps of a prototyping workshop

Step 1

In the beginning, we have a number of functions or initial solution scenarios that we would like to test. On the team, we ponder what functions are absolutely critical to a user. They are the functions we would like to integrate in the solution and test in the real world. As discussed in the previous chapters, prototypes exist in different manifestations and can be processed in different ways. What is important is that we implement something tangible and that an interaction with a potential user can come into being.

Step 2

The team thinks about which variant should be built.

Step 3

Now the team builds one or multiple prototypes. At this point, it is important to provide enough material for building the prototype.

Step 4

Performing the prototyping in several groups already allows us to obtain feedback from the others. A good way of obtaining feedback is through "green" or "red" feedback. The feedback is given in the form of "What I like about the prototype is..." (green feedback) or "I wish that the prototype..." (red feedback). This helps to maintain a positive basic mood and cultivate improvements.

Step 5

Based on the initial feedback, the prototypes as well as the way of presenting them are improved. It is important here to concentrate on the essential features and solutions.

Step 6

Before we go out and confront real users with our revised prototype, we carefully prepare our testing (see Chapter 1.10). One successful method is to go out in pairs for the prototype testing. One team member can ask the questions, and the other makes observations. After returning from the tests, all team members document and share their findings.

Step 7

Based on the findings, the prototypes are improved and/or some variants are discarded. If none of the prototypes work, it is useful to obtain more facts and customer needs and adapt the prototypes accordingly. The new variants of prototypes, in turn, serve for tests with potential users.

Prototyping workshop

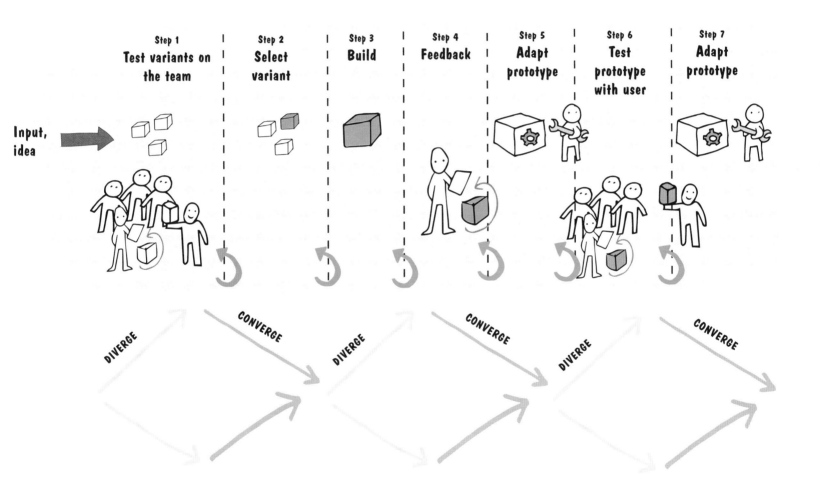

KEY LEARNINGS
Building prototypes

- When prototyping, start with the need of the persona and a trend in the market.
- Always build prototypes on the basis of the question of what is to be tested.
- Bear in mind that no offer has any intrinsic value. The value that customers ascribe to the offer is all that counts.
- Make sure that as many customers as possible ascribe value to the offer.
- Test the prototype as early as possible in the real world. Prototypes are assumptions that must be scrutinized.
- Use the material that is available to build prototypes.
- Create prototypes under time pressure. More time does not yield more results. Time boxing boosts the pressure to get results.
- Make sure that the objective and the maturity of the prototype match.
- Always schedule enough time for prototyping and testing, across the entire duration of the project.
- Involve at an early stage the project team members who will implement the prototype in the end.
- Apply "boxing and shelfing" to test a potential portfolio while prototyping.

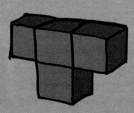

We always receive valuable feedback when we test prototypes with customers in the real world, namely potential users or in the environment of the users. Peter knows the importance of user and customer tests and tries to get out of his innovation and co-creation lab with his prototypes as early and often as possible. His current test of a prototype—an app for monitoring metabolic diseases with the option of receiving help from a team of doctors online—involves being out and about in Bahnhofstrasse in Zurich. Where else would he find the clientele for such an upscale and expensive managed service?

Test prototype

Understand > Observe > Define point of view > Ideate > Prototype > Test

Single-mindedly, Peter walks up to a pretty, elegantly attired lady in her mid-thirties, his prototype in hand. She is just leaving an exclusive shop for handbags and heads, loaded with bags, for her Bentley. She doesn't look sick or anything but Peter doesn't want to start the test with false assumptions. For Peter, the situation is clear: first offer help, come across as a likable fellow, and build up empathy. Peter is glad to carry the large shopping bags for the pretty woman. He asks if she would like to participate in the next big innovation, and two minutes later the two are enjoying a glass of Champagne for the "user test" in the bar kitty-corner from the exquisite shop.

The young lady likes the naive questions Peter asks with the attitude and behavior of a "greenhorn," so she talks a lot about herself but even more about the illnesses of the older gentlemen with whom she usually spends the warm summer nights in Monte Carlo. After one and a half bottles of Champagne—the mood is cheerful indeed—Priya happens to walk by the bar. At least now we know what triggered Peter's little marital crisis with Priya. Nonetheless, Peter has learned a lot about the application of his prototype, especially that there is no Wi-Fi coverage on yachts on the high seas, so getting online support from doctors there would be impossible.

Why is testing so important?

In tests with users, it is important to ask "why" in order to learn the real motivation, even if we think we know the answer. Our primary goal in a test interview is to learn, not to give reasons for or sell the prototype. This is why we don't explain (too early) how it works. We ask for stories and situations in which our potential customers might have needed the prototype. Whenever possible, we collect and analyze quantitative data to validate the qualitative results. This approach allowed Peter to learn a great deal about life in Monte Carlo and on the high seas.

Testing is an essential step in the design thinking process. Not infrequently, decisive change proposals appear during this phase that could enhance the quality of the end result substantially. In particular, the fresh views of people who were not involved in the development of the prototype and thus are much freer in their assessment can pay off quite well in the end. They can see prototypes through the eyes of a customer or user.

With a lot of empathy

HOW MIGHT WE...
design the test sequence?

A test can be broken down into four steps:

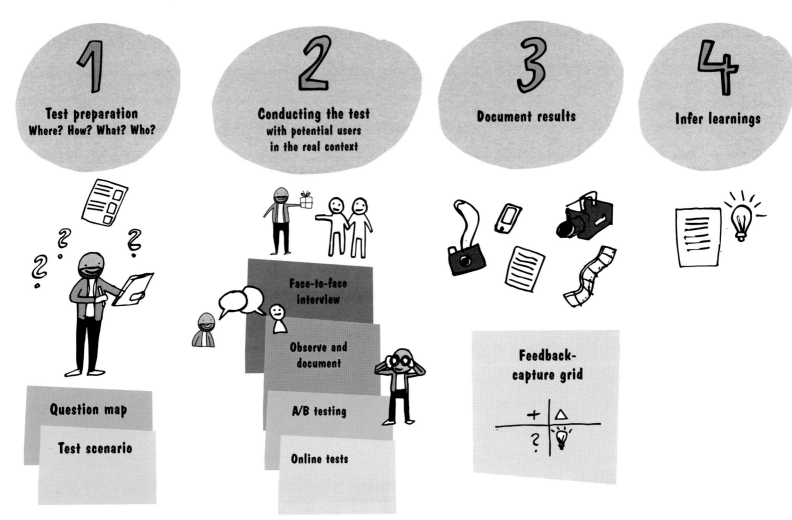

1

Test preparation
Where? How? What? Who?

Question map

Test scenario

2

Conducting the test
with potential users
in the real context

Face-to-face
interview

Observe and
document

A/B testing

Online tests

3

Document results

4

Infer learnings

Feedback-
capture grid

1. Test preparation

The best way to start is to define clear-cut learning goals or hypotheses that we want to test:

* What do we want to learn?
* What do we want to test?
* With whom do we want to conduct the test, and where?

In the end, the test should show what parts of an idea we should keep, what we should change, and what we should discard. In the early phases, the goal might also be to understand the problem. Before embarking on the actual test series with various users, an initial test with one person should be carried out to exclude any errors. We leave enough time to implement improvements after the first test prior to conducting more tests.

Define question maps

We formulate simple, clear, and open questions that we can explore in greater depth at the end. They should not be hypothetical but tie in to the real situation of the test person. We do not ask many questions, but rather focus on the core about which we want to gain insights. Courage and focus are important. Less important stuff can be omitted, so we don't overload the tests. We let the user talk about his experience. As the moderator, we can ask follow-ups when suitable; for example, "Tell us what you think while you do that."

Determine the test scenario

We reflect on the exact sequence of the test and the situation of the test person and describe it. We provide as much context as necessary and explain it as simply as possible. We let the user experience our prototype and deliberately refrain from explaining the thoughts and considerations behind our prototype. Particularly in phases of the design thinking process in which there are still many iterations, the issue is not, for instance, to find out how much the customer would be willing to pay for a product. Instead, we try to find out whether our idea matches the context and life of our user and, if so, how does it fit.

2. Conducting the test

It has been our experience that we achieve the best results when we test multiple ideas or variants of one idea that we have described as a scenario beforehand. This way, the feedback will be far more differentiated. If we only have one solution ready, the user's response to what he thinks about the idea might be rather vague. That usually doesn't get us very far in terms of our clarifications. When the user must undergo several different test arrangements, he can make comparisons, evaluate, and formulate his feedback far more precisely, such as what exactly he finds better or worse in one prototype than in the other. It has become second nature for us to test the prototype in a context, namely in its natural environment.

As mentioned, it is better to include more people for observation and documentation in the test, as per the motto, "Never go hunting alone." Those involved can take on different roles. For example:

The moderator:
As a moderator, we help the user to cross over from reality to the prototype situation and explain the context, so that the user has a better understanding of the scenario. In addition, it is our task as a moderator to pose the questions.

The actor:
As actors, we must take on certain roles in the scenario in order to create the right prototype experience—usually a service experience.

The observer:
The important task for observers is to watch in a focused way everything the user does in the situation. If we have only one observer on the team, it is best to film everything so we can look at the interaction together later in more detail.

Online tools can also be used for testing.

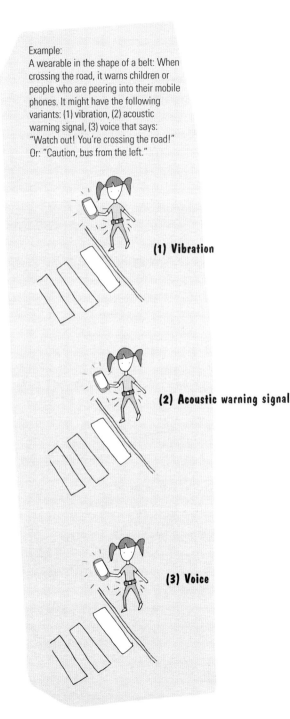

Example:
A wearable in the shape of a belt: When crossing the road, it warns children or people who are peering into their mobile phones. It might have the following variants: (1) vibration, (2) acoustic warning signal, (3) voice that says: "Watch out! You're crossing the road!" Or: "Caution, bus from the left."

(1) Vibration

(2) Acoustic warning signal

(3) Voice

3 Document results

In our experience, it is of vital importance to document the results. In so doing, we actively observe how the users use (and misuse!) what we have given them. We do not immediately correct what our test person is doing. Photos or video recordings are very suitable for documentation. We always ask the users for their permission. Digital tools make documentation easier, but be careful not to forget to use them. To elict richer answers, we probe with further questions. It's very important and often constitutes the most valuable part of the tests. Questions can be, for example, "Can you say more about how it feels to you?" "Why?" and "Show us why this would (not) work for you." Ideally, we answer questions with questions: "What do you think this button is for?" Resist the temptation to conduct a marketing or voice-of-the-customer survey!

The use of a feedback-capture grid has proven quite useful. It facilitates the documentation of feedback, either real-time or from presentations and prototypes. We use the grid to capture the feedback systematically and deliberately in four main areas.

- What do we like?
- What wishes do we have?
- What questions have cropped up?
- Which initial ideas and solutions have we found?

Filling in the four quadrants is pretty easy: We write each piece of user feedback in the suitable quadrant category.

FEEDBACK-CAPTURE GRID

Like
Things that somebody likes or that are worth mentioning

Wishes
Constructive criticism

Questions
cropping up during the experience

Ideas
generated during the experience or presentation

As an alternative, we can choose the following areas for the four quadrants: "I like…," "I wish…," "What if…," and "What is the benefit?"
This method can be easily applied to groups consisting of two to over 100 people. The simple structure helps to formulate constructive feedback.

Giving feedback is one thing, receiving feedback quite another. When we receive feedback, we should see it as a gift and express our gratitude. We listen to the feedback and do not have to answer in any way. In addition, we should avoid justifying ourselves and simply listen well. At the end, we ask again if haven't understand something or if something is still unclear to us.

4. Infer learnings

The insights serve to improve our prototypes and adapt the persona. Going through the iterations is crucial here; it contributes to constant learning.
The purpose of testing is to understand needs better and build up empathy. The approximation and constant improvement—as well as, again, failure and mistakes—achieve the learning effect. We all know the banal-sounding expression "fail fast—fail often." Early and frequent failure is indeed an important element of design thinking and contributes significantly to realizing market opportunities in the end. At the end of the testing, it is important to document both the findings and the test well and share both with the team.

EXPERT TIP
Carry out an A/B testing
with your prototype

One possibility of quantitative testing is to carry out an A/B comparison. It is especially suitable for simple prototypes and allows us to test two different versions of a landing page, for instance, or even two versions of an element such as a value proposition or test button. In the case of a Web site, the titles and descriptions of the offers, the text volume, style, promotion offers, length of forms, and boxes can be examined in an A/B test.

To achieve relevant test results, it is important for both versions to be tested concurrently or in tandem and within a predefined, appropriate time period. The final measurement and evaluation as to which version was more successful in the test and which one will be used in the real world must be done on the basis of clearly predefined criteria.

At an early stage of prototyping, we have the test person first experience variant A. Then we find out what the test person likes about it and what he would want changed. Then we repeat the procedure with variant B. Depending on the situation, we can also observe and question one test group about variant A and another about variant B.

Using a landing page, we can check the conversion rate directly in an A/B test by observing the reactions; we simply distribute the page views to version A and version B by means of an A/B testing tool. Only one variable at a time should be changed to find out why one variant is better liked. This A/B test shows clearly which Web site gets more registrations. Calculators are available to check the statistical relevance. If a Web site already exists and we want to test a new version B, we make sure that regular visitors don't get confused by making version B available only to new visitors.

The test can show a result in favor of A or B, respectively, or else no statistically relevant preference at all. Perhaps possibilities can be inferred from the test as to how to combine the best of the two variants.

ONLINE TESTING

What is better?

A

or

B

What digital tools can be used to test prototypes quickly?

An extremely simple and effective way of taking many users' feedback into account is the use of a Web-based tool. Recently, various Software-as-a-Service solutions have evolved, with which affordable, efficient and Web-based feedback can be obtained.

With the aid of such a tool, Peter quickly built up an internal feedback community consisting of employees of his company and selected external customers. "Friendly user test," a term frequently used in German-speaking countries, doesn't quite hit home. After all, the specific purpose of the test is to identify weaknesses in the design and get suggestions for improvement—which are not necessarily "friendly." The term "customer trial" used in English-speaking regions is a little better.

Peter has used such a tool for a customer trial several times already, and it has been helpful in his experience. It enables him to obtain feedback in relation to

- prototype variants,
- procedures, and
- images or links through URLs.

and to conduct A/B testing. The number of prototypes is unlimited. One great advantage of such a tool is that additional questions can be asked and there is a great deal of leeway in terms of the makeup of the community surveyed. The segmentation ensures that the feedback matches his needs optimally.

On the same day he sets up the tool, Peter receives some initial feedback. Within only two days, he can give a valid assessment of the prototype variants, based on which he can develop a new product function.

A tool-supported approach for testing feedback allows you to obtain structured feedback quickly and easily. When selecting the right tool, the following criteria should be kept in mind:

A. Does the tool offer the possibility for uploading various types of prototypes?

Example:

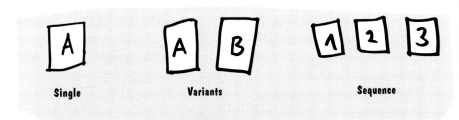

Single   Variants   Sequence

B. Is there a possibility for drawing up a scenario? This will give responding users the opportunity to see and understand the situation.

C. Does the tool enable us to ask predefined and open questions? It pays to spend much time on formulating the question, because it directly affects the feedback and its quality.

Examples of questions:

1. Evaluate the prototype with 1 star (poor) to 5 stars (really awesome).
2. What do you like about the prototype?
3. What would you change in the prototype?
4. . . .

D. Another key factor of success is the selection of the feedback community. Ideally, it should not be limited to one's own organization (university, company, etc.) but instead include the possibility of inviting additional, freely definable respondents for a survey.

Example:

It is useful when experts within an existing community have the possibility of selecting their field of expertise (e.g., channel marketing, big data analytics, accounting). This makes it easier in actual practice to obtain fast feedback, such as from the experts with respect to their expert knowledge.

The dedicated selection of technically accomplished community participants can boost the quality of the feedback, but you should always consider the feedback of nonexperts as well; because they are less profesionally blinkered, they often have a fresh viewpoint.

EXPERT TIP
How do we visualize prototypes
for tests in digital tools?

A prototype is the visualization of an idea. It can be a sketch, a photo, a storyboard, or a chart. Any offer can be visualized as a prototype early on and made available to a tester community for feedback:

Sketches

Wire frames

Drafts

Storyboards

Ideas

Videos

Web sites

Apps

Logos

HOW MIGHT WE...
conduct and document experiments in a structured way?

During the early phases of the innovation process, we frequently test several assumptions concurrently and learn on several levels. However, we recommend that you reflect before each test on what exactly you would like to learn and what the key question is. We also ask ourselves which assumptions we would like to test and how we can design the test scenario in such a way that the user can experience them.

Over the course of the further development of the product or service, we test our assumptions again and again and conduct experiments continuously. In the early phases of the innovation process, the prototypes are normally very simple. Often, several variables are tested at the same time. For the testing in later project stages, other types of experiments with customers (e.g., online tests, A/B testing, etc.) can be conducted. Here we usually focus on a single test variable or assumption.

It is of great importance that all tests/experiments be well defined. Documentation helps when tracing decisions later or showing an investor the success of an MVP. A simple experiment grid helps to structure the experiments and can be used to document the learning progress.

We want to learn as quickly and cost-effectively as possible; this is why we think about how the test (or the experiment) could be conducted in half the time and with half the resources. We ask ourselves whether there are variants that allow us to learn the same thing more quickly and economically.

The "experiment grid" helps define and document the tests/experiments:

In a first step, we describe the hypothesis we would like to test.

In a second step, the actual experiment is explained. The experiment can be a prototype we want to test with customers/users, an interview, a survey, and so forth.

In a third step, we define what we want to measure and which data should be collected. This can be a certain volume of positive feedback or just a specific value.

In a fourth step, we determine the criterion that shows whether we are on the right (or wrong) track.

In the next step, we carry out the experiment and document our learnings, such as with photos or videos.

At the end, we note the insights gained, the conclusions drawn, and what measures we will undertake. The tests/experiments must be well documented.

Experiment 1	Learnings 1
Step 1: Hypothesis We believe that. . . **Step 2:** Test To verify this, we will. . .	We have learned the following:
Step 3: Metrics And measure. . . **Step 4:** Criteria We are on the right track if. . .	Documentation of the test (e.g., photos)

KEY LEARNINGS
Test prototype

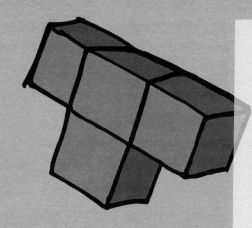

- Define scenarios and a clear goal before the test.
- Involve neutral people when conducting the test; namely, people who did not build the prototype.
- Ask simple and open questions in testing; never suggestive questions. Always ask "why" in order to find out the underlying motivation.
- Don't design the test to be too long. Concentrate on what is essential.
- Take along groups of stakeholders to the test (e.g., developers), so they can experience user feedback firsthand.
- Let the test persons think aloud and don't interrupt them. Don't try to influence them by steering them in one direction or selling the prototype as a great solution.
- Avoid the pitfall of relating too quickly or too much about how the prototype works.
- Document the tests, and always schedule enough time after a test to integrate the findings into a new prototype.
- Use Web-based tools for simple prototypes.
- Perform many qualitative tests with no more than five test persons in each one.

2. TRANSFORM ORGANIZATIONS

Our personas are always confronted with the question of where they should practice design thinking at their university or in their company. The premises of most companies and universities were neither planned nor staged as creative spaces, nor are they suitable for such use. The majority of them are filled to the gills with bulky furniture, thus blocking any creative energy. In particular the tables prompt people to work individually or else work on their laptop. In the best case, employees or students sit around a table, which encourages, at most, an exchange of ideas, but does not generate any shared common creativity.

The good news for Peter, Lilly, and Marc is that nearly every room that has plenty of natural light and space (preferably about 5 m²/55 sq ft per participant) can be quickly reshaped into a creative space. The goal is to gain as much freedom as possible for creativity to unfold. We do best by starting with redesigning the environment and implementing the first prototype of a creative space.

At the bank where he works, Jonny finds meeting rooms galore, but only very few of them have the necessary flexibility to foster creativity. He has brought up the need for such a space several times already. In the end, he succeeded in convincing his boss, while having lunch together, to venture the experiment of a creative space. The room he gets for it is not optimal, but the old coding machines that were stored there had to be disposed of sooner or later anyway.

How should the room look? What furniture do we need?

Start with emptying the room, because less is more in this case. Only in an empty space can something new evolve. We consider how many people are to be creative in it and put in one or two additional chairs, stackable ones if possible. Flexible and stackable material is better suited than inflexible and rigid stuff, because stackable furniture allows you to create even more space if the situation calls for it. The design of the space must take into account whether the creative space must accommodate a project team of 4 to 12 members working on a project for weeks or months, or 8 to 25 participants who sit over a topic only one or two days.

For feedback providers, the space can be furnished with additional stools or textile cubes. The textile cubes can also be used as seat stacks and staged beautifully. Feedback sessions last only a couple hours, not days, so simple seating accommodations are quite reasonable.

What material do we need for the workshop and the prototyping?

The next important thing is to think about the material to be used for building prototypes. You can use a caddy on wheels as a container, filling it with numerous multicolored whiteboard markers and Post-its in various colors and sizes as well as adhesive dots. Another variant is to keep everything in transparent boxes. Such boxes are particularly advisable if you often want to travel with your prototype material or switch rooms.

We've found it useful to provide some prototyping material as early as at the beginning of a workshop (e.g., playdough, Lego bricks, string, colored sheets of paper, cotton wool, pipe cleaners, etc.) and lay it out on the table in the room. Masking tape to hang flip charts is always useful—like other prototyping material, it can be purchased at any DIY store.

Depending on the size of the space, we still need one or several flip charts on rollers. If no flip charts are available, we can fasten individual sheets of the flip chart to the wall with nails, or simply paper the walls with individual sheets of paper and masking tape.

As an alternative to flip chart paper, large paper rolls can be used. Pieces can be either cut by hand or torn off with some integral device. The pieces can then be stuck on the wall with masking tape. From our experience, it's always good to have some extra flip chart paper. Nothing is more annoying and inhibiting to the creative flow than when we run out of basic material—this includes functioning whiteboard markers.

Usually, any smooth walls are suitable for working on flip chart paper and hanging it up. Should the walls be very uneven, several sheets on top of one another should be used in each case, so they can be written on legibly. As an alternative, you can work with Post-its in such a case, which can be written on prior to placing them on the flip charts.

If large paper surfaces are needed and XXL sheets are unavailable, we glue together any amount of flip chart pages with the masking tape on the back to build huge creative surfaces. Such empty creative spaces, even if they refer "only" to paper, are important because creative energy needs room to unfold. It goes without saying that the flip chart paper is used on the side with the squares.

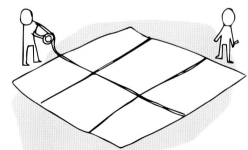

Writable walls, window panes, or glass walls are excellent for directly writing and painting on them with whiteboard markers.

If, for some reason, there isn't enough space on the walls, whiteboards and maybe pin boards on rollers, which can be moved around the room, are the right choice. Design thinking professionals use flexible whiteboard walls (in the HPI design) on rollers for their work.

If you're looking for a table in the creative space, it is extremely practical to use lightweight furniture that can be easily moved. Rollers are a plus.

With respect to the tables, choose a more organic, stimulating shape over a rigid rectangular one. The table should be placed free-standing in the space because, as described, all wall surfaces will be included in the creative work , so ensure there is sufficient clearance to work and move around the room.

Instead of on the table, we can simply put the required material on chairs or stools that are not needed. This uses less room, and we have more free space to move in. For the creative process, we don't arrange the chairs around a table, but instead distribute them freely in the room. When they don't sit stiffly at the table, participants stay more agile, both physically and mentally, which has an enormous influence on the creative process and the results.

If a coatrack is needed, it's best to use a stand that can quickly be moved for different settings and does not interfere with things in the space. As an alternative, you can put it outside the room. It is also import-ant that the bags and other luggage of the participants not be put on the floor along the walls but instead on top of or underneath unoccupied chairs. This is the only way to work on the walls free of hindrances and for the results to be presented and seen later.

WHITEBOARDS

HOW MIGHT WE...
further improve the creative space?

After the initial experience with the prototype of a creative space, we must now develop it further and improve it based on what we have learned.

1) What has worked well in the application? What would we like to have more of?

The next level of a professional creative space has whiteboards attached to the walls already. These are good for visualization. Important inputs and papers can be attached to them with magnets (extra-strong magnets for posters and heavier paper). Chairs are available in different colors, and stackable versions are preferred. The tables should have rollers, if possible, and should be foldable, so they're never in the way. Different working positions support the creative flow. As a supplement, tables on rollers can be very inspiring depending on the kind of workshop. Square measurements have proven useful; the tables used in the Design Space at the Stanford d.school have this shape. Four workshop participants can group around these tables, and enough space is left to sketch something or for prototyping.

More unusual material wouldn't go amiss for prototyping (Styrofoam, colored wool, wood, balloons, fabrics, cardboard, and the good old extensive collection of Lego bricks all find a new home here). Everything handicraft shops have on offer and that can be put into prototypes is usable. Lilly's favorite prototyping material is aluminum foil. Any shapes can be quickly created from it, and pieces can easily be made smaller without using scissors. There are no limits to the imagination—with time, you will realize that simple materials in particular have the potential for great prototypes.

2) How do we want to work in the future, and what helps our wishes come true?

With a more ample budget, the walls can be painted in colors that immediately create an inspiring environment. Colors such as orange, blue, or red are welcome; for example orange stands for creativity, flexibility, and agility, and blue for communication, inspiration, and clarity. Colors and patterns on usually barren floors are outstanding for inducing creativity. Carpets of all sorts, PVC, homey wood, or paint can be used, depending on the suitability of the subsurface.

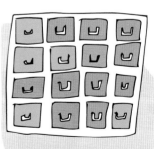

Treasure chest for materials

3) How do I find the right creative space for my organization?

Although you should set no limits to creativity, you should keep the industry, the type of enterprise, and the prevailing corporate culture in mind. The space can be enriched in a playful way with unusual and crazy things like rubber boats, hammocks, or shower curtains used as separation. Such objects can have an inspiring or consciously disruptive effect. This "disruptive" function of a creative space that dissolves or destroys what exists is quite conceivable in order to put things in motion. It's up to us and our sensitivity with respect to the other teams, our sponsor, or the decision makers to choose the right setting. Our tip: Begin with a low profile and observe the reactions of the environment carefully before you go too far with your creativity.

We can't tackle the challenge without some courage. It's not easy to change a work environment successfully. As with any innovation, you'll likely encounter resistance. Sometimes, such resistance points at real weaknesses in a concept; sometimes, people are simply suspicious. Any resistance must be taken seriously and accounted for in the implementation process.

A creative space can be designed jointly as part of a team development process. After all, the participants must feel comfortable and identify with their space. By the way, this is why employees often don't feel comfortable in stylish rooms: because their wishes and needs have not been taken into consideration sufficiently or at all.
Simple "goodies," such as active loudspeakers for a little music, might also be well received, because music can support the creative flow (e.g., soft music playing in the background during design sessions). The caddy can hold a coffee machine and an electric kettle for tea. Bottles of water, cups, and brain food such as nuts and dried fruits should be available in the room.

As an alternative to a small screen and a projector, teams can work with slightly larger screens if the available space allows it. Again, a version with rollers is recommended, so it can be pushed aside when not needed.

HOW MIGHT WE...
structure a prototyping workshop for the design of a creative space?

How exactly should we now proceed? First, we develop a common understanding of the idea, or the order and client. It goes without saying that we consider the scope, possible framework conditions, and any restrictions. This way, we arrive at an initial, roughly formulated design challenge that is open enough and does not contain any solution in its description.

We plan the estimated one and a half days for the workshop as follows:

1. As an input, we use the design brief and pictures of other creative spaces.
2. The procedure provides for a warmup in the morning, followed by individual brainstorming and several brainstorming sessions in connection with prompt translation into prototypes. The testing is done with employees in the cafeteria and coffee corners. At the end, the final prototype is presented to the decision makers.
3. As a result, we've now got two to three models of a prototype, which we will either continue to improve or order to be implemented.
4. Because the approach contains many elements of a design thinking cycle, the participants become familiar with it.

Design brief—creative space

The pace of change in business and the challenges entailed therein have become enormously complex. To cope with these developments, many companies and organizations have come to master tasks with more collaborative approaches to innovation.

Especially for creative working, workspace design is decisive for promoting communication and creativity. Companies such as Google, Apple, and Procter & Gamble have been pioneers in creating innovative and inspiring working atmospheres for flexible and individual collaboration.

The spatial environment is usually characterized by specific furnishings: flexible furniture, lots of space on the walls, the necessary tools and materials to visualize research impressions and new ideas, and suitable places to retreat to, in which ideas can be given their initial form.

We as a traditional banking house want to follow these developments as part of our digitization efforts. In a first attempt, we begin with a creative space for our co-creation workshops in the FinTech departments at our office in Singapore. In particular, business models, new business ecosystems, and first prototypes in conjunction with technologies such as blockchain are to emerge from these workshops.

Design challenge: How might a creative space look that gives us the flexibility to begin a cooperative innovation process with various stakeholders (internal/external), taking into account our values and our brand as a branch of a traditional French financial institution in Southeast Asia?

How might the workshop agenda for the two days look?

Prototyping workshop, model creative space		
• Purpose • Common understanding • Develop first prototypes • Obtain feedback from stakeholders		
Input	**Sequence**	**Output**
• Order • Framework conditions • Design challenge • Pictures of creative spaces • Material for prototype	• Warmup • Common understanding • Brainstorming: "what if?" • Interviews • Ideate • Create prototype • Test + feedback • Develop prototype further • Pitch before the jury • Next steps	• Refined design challenge • 2–3 prototypes (models) of creative space
Resources		
Catering, tables, chairs, pin boards, flip charts, blank walls, . . .	Timer, prototyping material, Post-its, pens. . .	Team, facilitator, jury, . . .

We need an environment that is familiar to us and with which we can identify and in which we feel comfortable. The designing of such an environment is essentially about four elements: the **place,** the **people,** the **process,** and the **meaningfulness** of the work. The **work environment** has become one of the most important instruments for a company to retain the best talents and high performers. Does anybody today want to work in an office that radiates the faded charm of bygone days and was probably squeezed down to the last square foot?

Companies like Google or IDEO are good examples of shaped work environments; Apple's new headquarters in Cupertino, California, initiated by Steve Jobs in 2011, is very inspiring. The building was consciously planned in a natural environment, surrounded by a forest. Jobs' vision was to create the best company building in the world. The building stands for the future and resembles a spaceship. The new headquarters was to take into account all sorts of desirable things for people.

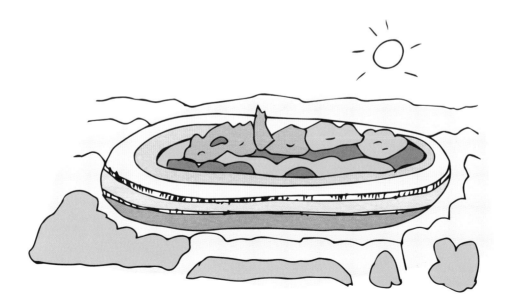

The process and the way the work is done likewise have a great impact on the results. For one, the focus is on the type of activities we must perform; second, on the interactions of the people among one another and their influence on the course of the project. Bear in mind here that the work process itself is in a constant interaction with the environment and the people involved.

The meaningfulness of what we do as motivators is often underestimated. Companies often lack a clear strategy from which the teams could deduce whether their activities aim at something greater. Surprisingly, the majority of companies have a hard time defining the "why." Especially for the much-cited millennial generation, meaningfulness is a key criterion for choosing an employer. There is no question that a meaningful activity boosts motivation. This applies to all of us. We will come back to the theme in Chapter 2.6.
In many cases, the management of a company is unable to cope with new, rapidly changing framework conditions (e.g., digitization). This uncertainty leads to greater aimless action in the company, but little work is done toward a specific goal or a defined market position.

In Chapter 3.6, we will discuss the question of how to deal with such uncertainties and present approaches and methods we can use, for example, to initiate and successfully implement digital transformation.

Meaningful work!

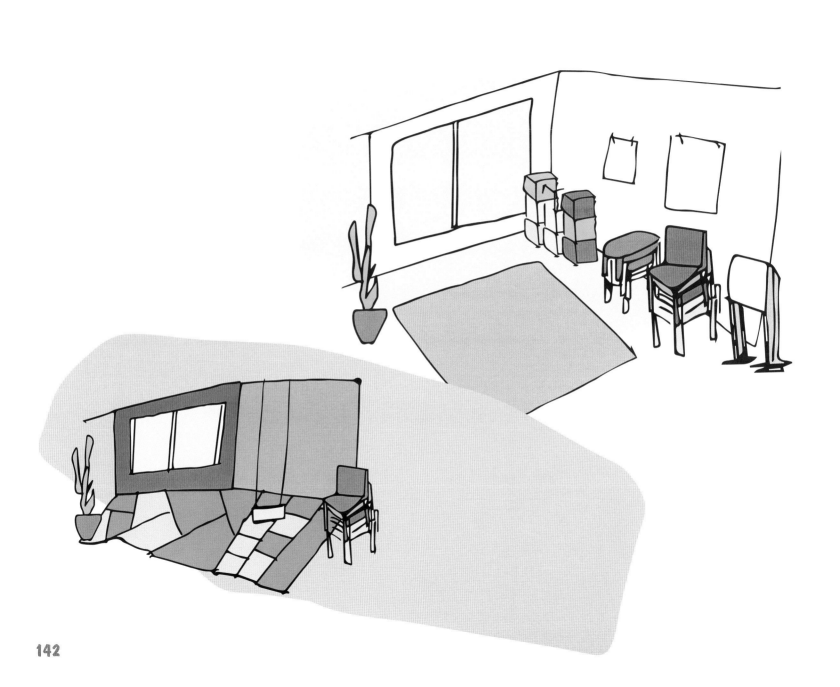

KEY LEARNINGS
Design a creative atmosphere

- Make sure that the environment is not overloaded. Less is more: Creativity mainly requires a great deal of freedom and space.
- Use rolling and stackable material. It offers the greatest degree of flexibility in the versatile use of a creative space.
- Stack tables efficiently by putting one on top of another, with the tabletop to the front, so it can be used for sticking inputs and results onto it.
- Try out as many different creative spaces with the teams as possible to find out what influences collaboration and results positively.
- Conduct an initial workshop with members of the management in an external creative space so as to induce enthusiasm in them for the positive impact of an inspiring environment.
- Change the space, the location, and the environment as often as needed. Avoid spaces that are fraught with boring memories.
- Design not only the space but also the work environment. This includes the processes along with the meaningfulness of the work.

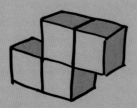

Peter collaborates with different teams in his projects. To be successful, he emphasizes two dimensions: The team members must have in-depth technical knowledge as well as a broad general knowledge. For Lilly's students, it is a wonderful feeling when they have finally advanced a step with their own question and gotten out of a deadlock. It often occurs because the participants have asked others for advice. Regarding the same problem from a different perspective often helps find a way out of a dead end.

With many problem statements, there are limits to how much your own skills can contribute to the solution of the problem. The reason for this is usually a lack of knowhow and experience in a specific subject area. No later than this point, the design team must consult an expert to get ahead. Frequently, it so happens that the expert begins his or her own work way before the actual topic is discussed and poses critical questions instead of simply tying in to their area of expertise. As a consequence, the things that were developed so far assume a new quality because they are suddenly considered using a holistic approach and not from a limited perspective.

The principle of iteration is, as you will know by now, a crucial element in design thinking. Take one step back, do another lap; it helps you get closer to an ever better product, which corresponds to and meets customers' needs. However, the most important thing is that we learn and iterate at a fast pace. This, in turn, only works when the questions are asked—and challenged—as early as possible and the things developed so far are looked at from a different perspective. The most promising way to achieve this is the exchange with potential users and on the team, which consists of different experts with in-depth and broadly based knowledge.

Holistic way of thinking

The principle of iteration

What characterizes an interdisciplinary team?

In very general terms, **interdisciplinary** means comprising several disciplines. On interdisciplinary teams, ideas are produced on a collective basis. In the end, everybody feels responsible for the overall solution. A methodological and conceptual exchange of ideas takes place on the way to the overall solution.

In comparison with **multidisciplinary teams**, they have the advantage in that, at the end, everybody stands behind the commonly created product or service—a factor of success that multidisciplinary teams, for example, cannot afford. On a multidisciplinary team, every member is an expert who advocates his or her specialization. The solution is often a compromise.

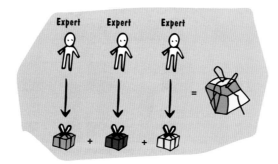

As mentioned, Peter wants to rely on the in-depth and broadly based knowledge of his teams. This idea is based on the principle of so-called T-shaped people—people who have skills and knowledge that are both deep and wide. The visual profile of skills is like a "T." The concept was developed by Dorothy Leonard-Barton.

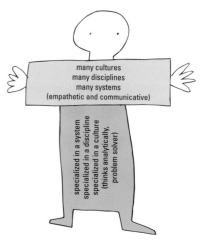

The vertical bar of this "T" profile stands for the respective specialist skills that someone has acquired in his training and that are required for the individual steps in the design thinking process and the implementation project. A psychologist, for example, brings along experience and methodological knowledge to the "Understand" phase.

The horizontal bar is defined by two characteristics. One is empathy: This person is able to take up somebody else's perspective while looking past his own. The other is the ability to collaborate as well as interface expertise: T-shaped people are open; interested in other perspectives and topics; and curious about other people, environments, and disciplines. The better the understanding of the way others think and work, the faster and greater the common progress and success in the design thinking process.

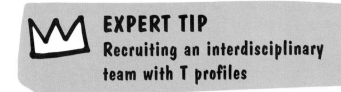
Lilly still plays with the idea of founding a consultancy firm for design thinking. She would need to recruit her future colleagues against a T-shaped profile in order to cope with the professional challenges and be able to work as a team, on which all have the same social skills. It is probably easier to find people with specialist skills for the individual process steps—corresponding to an I profile—than those who have both forms of knowledge.

A good sign for the ability to collaborate, from the horizontal bar, is when people during the interview talk not only about themselves but also emphasize what they have learned from others and how valuable the collaboration was for the common project.

Quite specifically, it can help to have the potential candidates create their own **T profile**. The way somebody fills out the profile yields a lot of information on his way of thinking and expressing himself. At the same time, it shows how somebody interprets the requirements for collaboration and presents himself in this respect.

If you want to take the time and experience potential team members in actual practice, you can hold something like a **design thinking boot camp**. It can serve several purposes: For one, it is a quick and easy way for candidates to experience design thinking and its individual steps in practice and judge for themselves if they want to collaborate in this way. Second, those who assemble the team quickly get an idea of the specialist and social skills of future team members.

Interdisciplinary teams have many advantages, which, among other things, lead to a better-quality result within a shorter time. At the same time, the complexity of the collaboration increases with this approach compared to an individual way of working, without iterative agreements. The complexity can be reduced using a few simple rules, on which the entire team should agree from the outset if the collaboration is to be successful. Some of them already comply with the principles of design thinking anyway, but it has proven to be of value for the team to reflect on them consciously again.

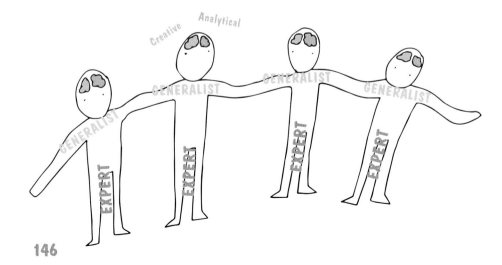

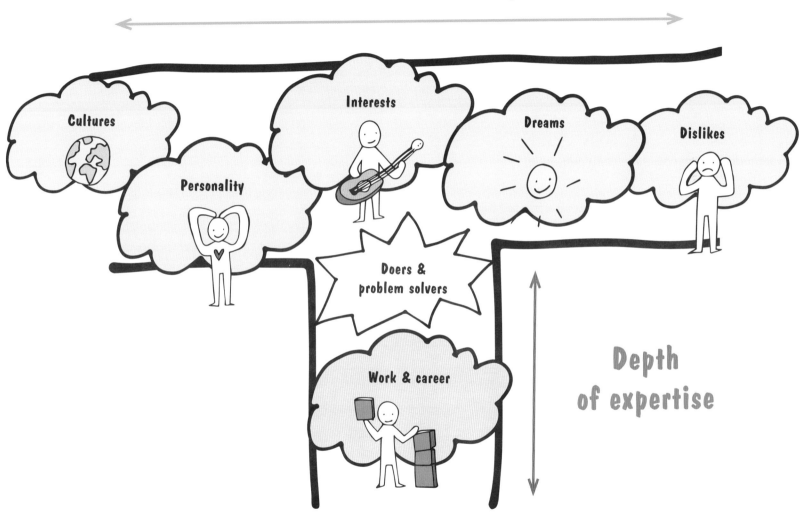

Breadth of knowledge

Cultures

Personality

Interests

Dreams

Dislikes

Doers & problem solvers

Work & career

Depth of expertise

147

HOW MIGHT WE...
formulate simple rules for collaboration on interdisciplinary teams?

The sooner the strengths of each individual team member can be experienced, the more interdisciplinary teams are able to benefit from the skills of the others in order to achieve the common goal. Putting teams together with people not only from various disciplines and departments, but also from different hierarchical levels, has proven to be of particular help in practice. Besides the exchange of specialist knowledge and methodological expertise, it also gives the team access to a broad knowledge and the necessary problem-solving skills. As a by-product, the new interdisciplinary approach will spread faster and transversally throughout the whole company, so this type of collaboration will be better understood on all levels.

Six simple rules for a successful interdisciplinary team:

1. The team has a common vision that must be fulfilled as a team. In the best case scenario, it will be an answer to the "How might we . . ." question.

2. Every step in the design thinking process is led by the respective expert (vertical bar in the T profile) on the team, who suggests a clear direction and tried-and-tested methods while offering support in the implementation.

3. The team has adopted common values. They have been developed together and are visible by everybody at all times. The brainstorming rules, for example, are a good basis on which collaboration on the team can be adapted and expanded.

4. There is an atmosphere of trust in which everybody has respect for and accepts the experience of the next person—at least when the role of expert is taken on.

5. Only those who know the expectations and to what extent they can be met can become better. The more comprehensible the feedback of the team is, the more specific will be the way in which the entire team and its individual members become better and ultimately act in common.

6. Shared common processes and quality standards are determined, so that everybody always knows the procedure and the necessary requirements for the desired result and can orient themselves to them.

With respect to the founding of her company, Lilly is already thinking about the future. She is not satisfied having "only" T-shaped people on her team; she also thinks about how employees should develop. Her idea of the ideal company size is a team of 15 to 25 employees, in which technical skills are represented multiple times. This would allow her to process several projects at the same time. To be able to mix employees and constantly put together new teams that mutually inspire each other, the professional and human basis just described is decisive. Within the framework of a learning organization, Lilly finds it important that her employees constantly develop: from T-shaped people to Pi-shaped people.

This corresponds to the profile of an adaptive employee who develops further in addition to his specialization. Such an employee is not only able, like in the T profile, to steep himself in the discipline of his colleagues and understand it; he also has the ability to respond to challenges of everyday working life and educate himself accordingly. In this way, he can assume multiple roles, which usually are linked to one another in terms of content: for example, business analyst and UX designer or software developer and support employee. Within the company, such a profile contributes to increased flexibility in the composition of the teams, something especially relevant to smaller companies with limited resources and a quickly fluctuating order situation.

Two things are vital for a successful transformation:

1. Identify gaps on the team and the development potentials of individual employees.
2. Draw up a training schedule to close gaps and promote employees.

In step 1, the company management and team leaders identify gaps and potentials and discuss personal interests and associated development paths with the employees. Subsequently, Lilly should create a training schedule with her employees and the team leaders that is attuned to both corporate and personal goals and verifiable by means of defined milestones.

Breadth of knowledge

Depth of expertise

Breadth of knowledge

Depth of expertise in one's own core competency

Depth of expertise in leadership

In addition to the basic professional and human skills and development schedules, one component is of particular importance to Lilly: harmonious togetherness, a team in which everybody can rely on everybody else. What is important to Lilly is that her employees like coming to work, feel accepted and safe there, and fall softly if they fall. The technical and professional expertise that is represented multiple times on the team, a learning organization, and empathetic employees constitute a good solid basis for a so-called **"U-shaped team."** This form of a team is also crucial for a high level of effectiveness because safety and security mean productivity.

The name can be inferred from the analogy of established systems, which, after a disruption, return to an equilibrium and are more stable than they were before. Systems that break at the first sign of friction are the opposite.

U-shaped teams help one another, are there for one another, even if one member has a bad hair day and is not able to deliver his usual performance; these are teams in which you can make a mistake without having to be afraid of losing your job.

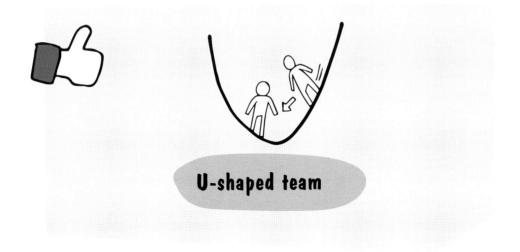

U-shaped team

Lilly's motto: People who feel safe, secure, and comfortable, who are supported and appreciated—with all their rough edges, warts and all—are highly motivated to deliver a great performance.

The application of the design thinking mindset and the associated methods with an interdisciplinary team are key factors of success. From a business point of view, the principles of T-shaped people developing into Pi-shaped people should always be borne in mind—on the basis of a U-shaped team.

In the section on the working environment, we already discussed the key factor of success for teams of needing the conviction that their task and their goal are meaningful and useful. The idea of a "team of teams" will be examined in Chapter 3.4, which highlights the successful implementation of market op-portunities. We would like to note at this point, though, that the personal relationships of and networking between individual people beyond the teams must be seen as decisive in the success of teams.

non—U-shaped team

The ideal is that we can equip our teams with people who have different preferences in their thinking; this is how we get high-performance teams in the end. The myth that the left hemisphere of the brain is responsible for analytical thinking and the right hemisphere for holistic thinking is widespread. But because our brain is anything but well organized, such a simple division into two halves is quite wrong. Tendencies can indeed be identified. Our understanding of numbers, for instance, or our ability to think spatially and recognize faces, is more localized in the right half of the brain. Other capabilities, such as recognizing details or capturing small time intervals, are localized more in the left half of the brain.

There are models that try to capture the brain as a whole and determine preferences in thinking. One example of this is the Whole Brain® model (HBDI®) that breaks down our brain into four physiological brain structures. The model consists not only of the left and right mode but also involves the cerebral and limbic mode. This view allows us to differentiate more styles of thinking such as cognitive and intellectual, which are ascribed to the cerebral hemispheres, and structured and emotional, which describe the limbic preferences. In our experience, it can be quite valuable to swap the thinking preferences on the team; individual tasks are assigned to the respective people in the relevant phases during the design process. Ultimately, better solutions are achieved this way. It also helps us when communicating ideas and concepts to decision makers and telling the story of new products and services we want to launch on the market.

If we have little time and want to learn something quickly about the participants in a workshop, we still use the model of the two halves of the brain. It helps to classify the participants roughly into the categories of "analytical/systematic" versus "intuitive/iterative." We have learned that the combination of the various thinking models and thinking preferences is essential for a successful design thinking project.

Metaphorical model (HBDI®)

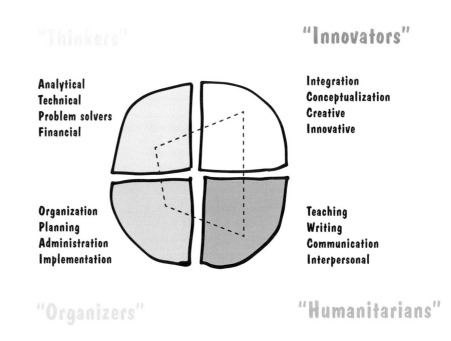

"Thinkers"

"Innovators"

Analytical
Technical
Problem solvers
Financial

Integration
Conceptualization
Creative
Innovative

Organization
Planning
Administration
Implementation

Teaching
Writing
Communication
Interpersonal

"Organizers"

"Humanitarians"

Marc and his team are already quite well set up for their start-up: Innovators such as Marc, women with a business sense like Beatrice, makers and shakers such as Vadim, and team members like Stephan and Alex, who actively control the business development.

In our experience, teams are most effective when they have one team member each with a strong manifestation of one of the quadrants in the metaphorical model.

Marc would like to shape the future. He has a great vision for the health care system of the future, in which patients decide autonomously which information goes into their electronic patient record. In a first step, he focuses on a few functions of his idea that he will implement on the blockchain and with the relevant actors in the ecosystem. Marc's HBDI® profile is characterized by a focus on the yellow and blue quadrants.

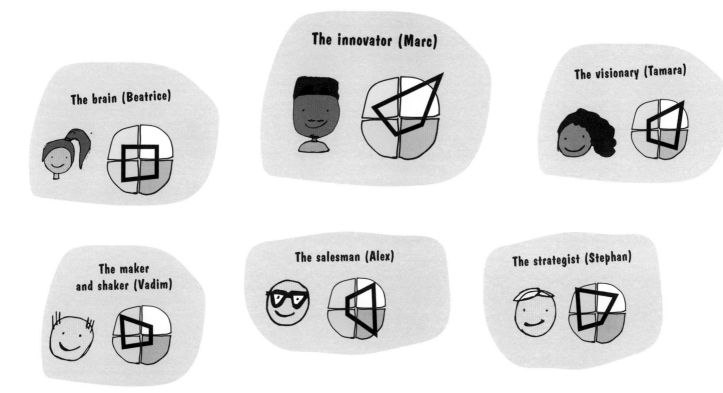

For the set-up and growth phase of their start-up, other skills that Marc and his team already possess are necessary alongside Marc's great vision and his programming skills. By way of example, we can place the respective HBDI® profiles on an S curve and put Marc's team members across the growth phase of a company.

Rational decisions must be made and measures defined for the next scaling step and for the transformation of the company from a start-up into a mid-sized enterprise. Above all, team members with strong organizational ("green") and interpersonal ("red") characteristics are necessary for this. Both are currently underrepresented on Marc's start-up team. When expanding his team, Marc should therefore make sure that this missing expertise and these capabilities are included on the team.

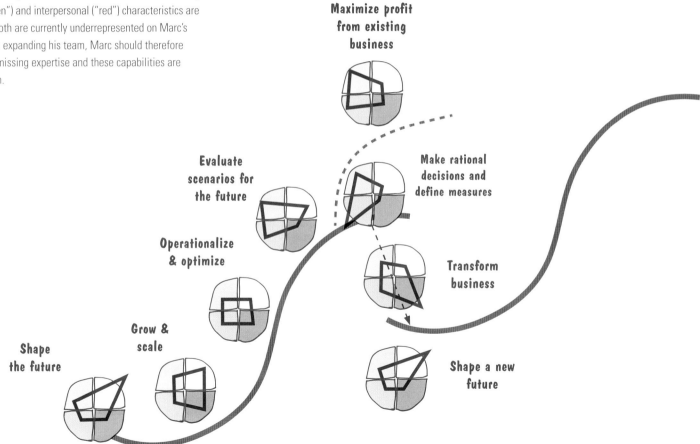

Maximize profit
from existing
business

Evaluate
scenarios for
the future

Make rational
decisions and
define measures

Operationalize
& optimize

Transform
business

Grow &
scale

Shape
the future

Shape a new
future

Using the teams of teams concept, the respective skills in larger organizations can be utilized for all design challenges. This will pay off once Marc's company grows and individual squads work on different functionalities for patients and for involving other actors in the ecosystem.

Marc works with the six principles from the "Connect 2 Value" framework (Lewrick & Link) in order to develop his start-up successfully.

The "Connect 2 Value" framework is based on three levels:
- Connect knowledge to value
- Connect talent to value
- Connect systems to value

It combines the design thinking mindset with the core aspects of a human-centered approach, co-creation, and value creation as well as with strategic foresight for the definition of a clear vision for the enterprise.

It also combines the talents of people in the company so their knowledge and skills can be unfolded to their full potential through the connections in the business ecosystem, and ensures that top talents are deployed in those places where they can create the greatest value for the company. Relevant internal and external needs are satisfied by the human-centered culture and its positive energy. Internal teams and expanded teams from the business ecosystem create value together.

1. People: Attract the right talents

Marc does not just consider the skills, the T profile, and the HBDI® model but also invites potential candidates directly to his workshops to find out whether they generate a positive energy and mood on the team and live the design thinking mindset.

2. Realize team and network effects

The composition of the teams and a goal-oriented collaboration, internally and with external partners such as in the business ecosystem, are key factors of success. Marc sees the human beings behind the companies in the business ecosystem as team members, too.

3. Shape for growth & scale

From the outset, Marc co-designs the business ecosystem. He creates a win-win situation for all actors. He uses technologies and platforms for scaling, which also make the realization of data-driven innovations and business models possible.

4. Create the magic

For Marc, leadership means to bring the magic to the team and make possible the impossible for them. It includes the creation and communication of a business vision, which helps to encourage the teams to fulfill their mission and act in an entrepreneurial way.

5. Create a new mindset

Marc is aware that positive energy is the elixir and driver of outstanding achievements and motivation. An open feedback culture, for instance, allows the top talents to contribute their skills optimally. Ideas and concepts are implemented quickly, and failure is part of a positive learning effect.

6. Make it happen

With the support of the top talents, the business ecosystem, the right mindset, and the right processes, Marc implements the concepts—fast and agile. In so doing, Marc assigns the top talents and resources to those activities that generate the greatest value. He consciously uses external skills and platforms of actors in the ecosystem for the realization.

CONNECT 2 VALUE FRAMEWORK

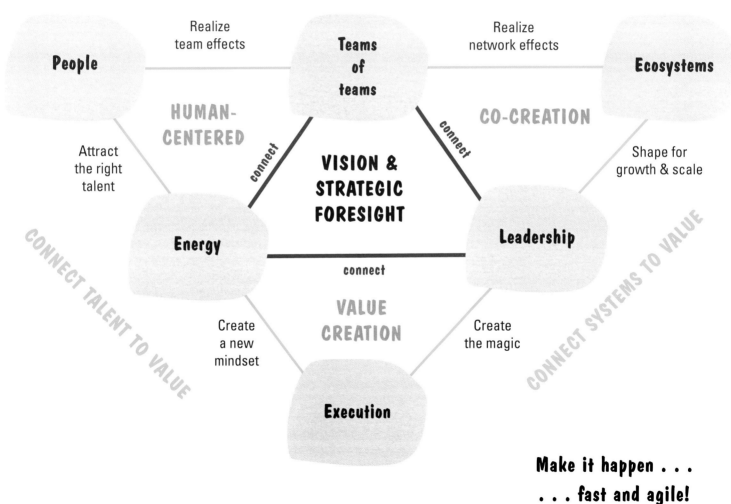

CONNECT KNOWLEDGE TO VALUE

People — Realize team effects — Teams of teams — Realize network effects — Ecosystems

HUMAN-CENTERED

CO-CREATION

Attract the right talent

Shape for growth & scale

connect — VISION & STRATEGIC FORESIGHT — connect

Energy — connect — Leadership

CONNECT TALENT TO VALUE

CONNECT SYSTEMS TO VALUE

VALUE CREATION

Create a new mindset

Create the magic

Execution

Make it happen . . .
. . . fast and agile!

KEY LEARNINGS
Put together interdisciplinary teams

- Put together interdisciplinary teams of T-shaped and Pi-shaped members.
- Let the team members draw up their own T-shaped profiles and present them to one another.
- Develop a common vision of collaboration and adopt common values and rules.
- Create an atmosphere of trust and respect on the team.
- Involve all disciplines in the project work. Make sure that the technically relevant perspectives and personality-related views are represented in equal measure.
- Visualize the different thinking preferences, such as with HBDI®—this improves mutual understanding.
- Take advantage of the heterogeneity of the teams in the form of differing approaches, thinking preferences, and background knowledge in a targeted manner in order to promote creativity.
- Identify weaknesses on the team. Define measures and develop team members' collaboration skills step by step.
- Use the "Connect 2 Value" framework to implement your projects—fast and agile.

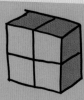
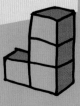

2.3 How to visualize ideas and stories

"But we don't know how to draw . . .!" This is a statement Lilly often hears from her students and attendees of her design thinking workshops. Visualizing is not the same as drawing, Lilly tries to explain while encouraging students to think with the pen. Visualization is a powerful tool for us to bring abstract information and interconnections as well as data, processes, and strategies into a graphical (i.e., visual) form. In design thinking and in workshop moderation, we can use visualizations in various phases of the process. Visualizations help convey themes and problems comprehensibly to our teams and users. We process visualized content faster, understand it better, and remember it longer.

With the help of quick sketches and visualizations, we can pursue different goals:

- Outline many ideas as part of a brainstorming session.
- Develop a common understanding.
- Make abstract things tangible.
- Create a dialog by collaborative drawing.
- Find surprising solutions with sketches.
- Draw the function of a prototype.
- Sketch a customer experience chain and bring it to life.
- Lighten the mood and make content more interesting.
- Shape the story in a lively way like we do here in the *Playbook*.

AND WHICH PEN
ARE YOU??

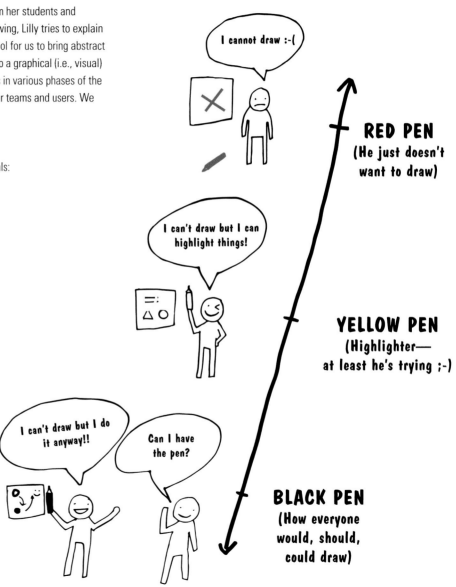

158

When we recall our childhoods, making drawings was something we did every day. So, at one time, we were all courageous. When we dig out the drawings from our childhood, we recognize they are reduced to the essential. They are usually simple and make use of repetitive elements. We can adopt this basic idea for visualizing, because visualization is neither art nor artistic drawing and it doesn't have to be beautiful, either. The point is the simple communication and quick transport of content.

Some examples of children's drawings that we all understand:

We recognize a house, milk, a mouse, Christmas, a fish, and a dog. If we were able to draw as children, we can still do it today!

What is the difference between visualization and artistic drawing?

If we want to visualize the function of the heart, for instance, we don't have to depict the exact anatomy photorealistically. It suffices to show the most important elements schematically: where the heart is located and what its function is. Visualizations should be created fast and to the point.

We present an example of blood circulation in a very simple depiction, so it's a joy to understand the content.

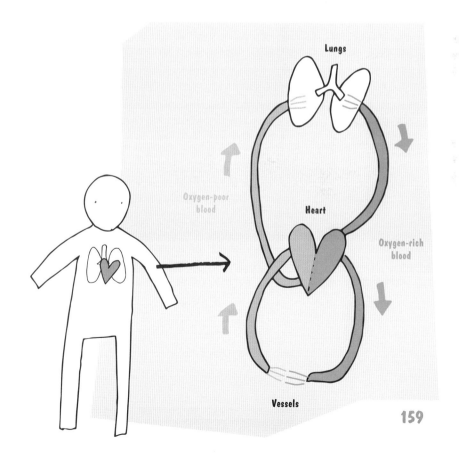

159

What makes a good visualization?

Good visualizations direct the eye to what is essential. The trick here is to leave out what's nonessential. This means no art in the sense of embellishing, decorating, or designing beautifully. Our goal must be to be as vivid, lifelike, and specific as possible.

Four properties are crucial in the creation of visualizations:

- We focus on what's important—and leave out all that's unnecessary.
- We are specific—we don't create vague drawings.
- We make our pictures comprehensible—and are able to make linkages to the content.
- We kindle interest—it is delightful to look at the pictures.

Lilly motivates her students to use visualization for specific design challenges. How would a child-friendly can opener look, for instance? The large number of sketches quickly show which lines are identified and what symbolism is clear or irritating. Another exercise is to put into a picture what we were just speaking about with somebody on the phone. Comics are ideal if you seek inspiration for facial expressions (e.g., Calvin & Hobbes or similar comics collections). Competitions on the Web can be motivating if we want to become better. Usually, they are about visualizing individual concepts, which are then published on Facebook or Instagram.

Other applications of visualizations include:

1. **Creative thinking:** We outline our ideas and show interconnections (visual thinking).
2. **Presentations:** We want to convey our knowledge to others comprehensibly (presenting).
3. **Documentation:** We record the knowledge of the group (graphic recording).
4. **Exploring:** We learn in common by presenting and documenting (visual facilitating).

In our experience, things get most exciting when everybody in the group begins to visualize and depict what they think in pictures; this way, a common shared image or a vision can emerge.

1 Creative thinking

2 Presentations

3 Documentation

4 Exploring

HOW MIGHT WE...
use the key design elements for visualization?

In principle, as mentioned, various elements of content are required for visualization. The graphic depiction is usually the composition of different design elements. They include:
text (1), graphic elements (2), icons and symbols (3), figures and emotions (4), and color (5).

We can depict all sorts of things with these design elements: ideas, stories, processes, diagrams, and so on.

Text (1)
When we use text, we should keep a few things in mind:

- We pay attention to legibility and select basic font types!
- We write from left to right, starting at the top left.
- We leave enough space between letters and lines.
- We formulate short and simple sentences and use familiar words.
- We provide structure by means of headings and visual blocks.
- We use stimuli such as different styles and colors.

Peter's handwriting is terrible, but he found his own technique of legibly writing on Post-its and adding to it through visualization. At first, he thought, if nobody can guess what the writing implies, chances are better that the idea will be shared with others by including the sketch. Peter has attended a scribble course where he learned that you don't have to be a good artist with fine handwriting to visualize well.

Different line widths and styles

LIGHT   CONDENSED   OBLIQUE
REGULAR   EXTENDED
BOLD   OUTLINE

I WRITE IN PRINT

Graphical elements (2)

Simple graphical elements are containers, folders, lines, and arrows.
They help to establish interconnections and order. Write the text first, then delimit or connect it with the graphical element.

Examples of containers, folders, lines, and arrows

Figures and emotions (4)

Because design thinking is always based on people, we've learned it is useful to be able to draw stick figures, round figures, and those shaped like stars. If you manage to draw metaphorical elements, you bring more character and emotion to the visualizations.
We can easily depict figures and their emotions. Again, it is important only to draw what's absolutely necessary.

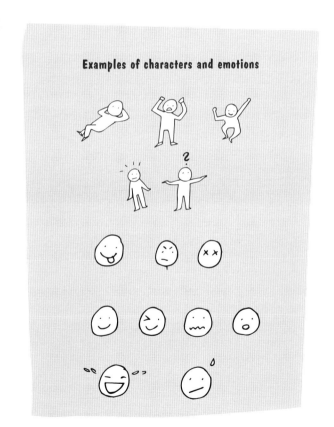

Icons and symbols (3)

The use of icons and symbols makes visualization more intriguing. Icons and symbols are visual abbreviations. The icon is a reduced picture of an object. The rule for icons is: **The simpler, the better!** The icons are not there to replace or decorate the text.
Symbols are signs that have no resemblance to the actual object and stand for something.

Examples of icons

Coffee Letter

Shopping cart Sound

Examples of symbols

Idea First aid

Radioactive Love

Examples of characters and emotions

Color (5)

For a composition, we recommend using only a few different colors. Colors are used to highlight or illustrate interconnections. Too much color can be confusing.

Visualization works best when the important stuff is highlighted, such as by coloring or adding a border, underlining, or cross-hatching. We use the space generously. Again, the motto is: Less is more!

With the design elements presented, we can create compositions or even charts.

Charts can help us compare numbers, sequences, size ratios, processes, and structures. Together with the main lines and connections, they are quite compelling.

When using charts, it is important to keep these points in mind:

- We use bar charts or pie charts for the transparent depiction of numbers or figures.
- To depict structures and processes, we use visualization by means of organizational charts (e.g., arrows).
- Portfolio charts help us to depict relative sizes and position the individual objects among one another.

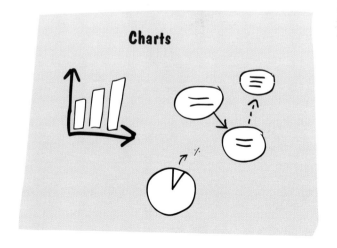

How do I prepare a visualization optimally?

After becoming familiar with the design elements, we can now think about planning a visualization.

For beginners, it is often difficult to create a good visualization spontaneously. It helps to concern oneself first with the core messages to be conveyed. Think about what symbols are important and which icons would fit.

The four WH questions in the preparation phase:

- Content   What do we want to depict?
- Goal   What is the purpose of the depiction?
- Target group   Whom to we want to give information to?
- Medium   What tools do we use?

Where can we apply visualizations?

We can carry out visualizations on all media: from the flip chart to a Post-it, on the iPad or just in our notebook. A good pen is half the battle. Poor pens will produce poor results, which will frustrate us right at the beginning.

We have various possibilities for drawing on a flip chart. Depending on the motivation and the goal, the theme can be drawn at the top or in the center.

Using high-quality pens and checking that they all work well will pay off. We should try to write in big letters, at least 3 to 3.5 cm/1.5 inches, or even bigger. Be careful to hold the pen correctly: set it down squarely and write with the wide side.
In addition, we use frames, graphical bullet characters, and simple symbols. If chalk is at hand, we can add color afterward.
The goal is to structure the flip chart in a meaningful way and create an attractive composition:

In design thinking, we often have very large walls with many visualizations, such as life-size personas, insights from customer interviews, images from surveys together with sketches of ideas, and customer experience chains. In the end, each team structures the space on its own. The facilitator can give hints, so that, at the end, the "journey" is comprehensible even to outsiders. Often, large Post-its are sufficient to show where the participants stand in the process. Our experience has been good with connecting elements and structuring grids, which, in conjunction with lines and arrows, result in an overall picture.

LESS IS MORE!

So much text

The most important stuff

If you want to learn how to visualize, you'll find much on offer in books and training courses that are dedicated to presenting one specific approach as the only valid one. We, by contrast, rely on our design thinking mindset. Our motto is: Just do it! Now.

What does it mean?

We have the courage to draw our lines, our own circles and ellipses, even if they look terribly crooked sometimes. Because as soon as we have practiced a little, we might be able to get a handle on the crookedness and turn it into our trademark. We are creative with regard to our own symbols and icons; we aren't just referring to a general library of icons and symbols. It is similar to language: "The meaning of a word is its use in the language," said Ludwig Wittgenstein. The same is true of symbols and icons—the context and visual cultures in which we visualize play a major role.

So just do it and begin immediately sketching and visualizing—everywhere and every day, because this is the only way to immerse ourselves in the new language. Although we do live in a world of images, we have a hard time imagining even very simple things, let alone complex stories and themes. This is why we must not only train the actual activity of visualizing but also our imagination, thinking in pictures and images! Every day!

KEY LEARNINGS
Visualize ideas

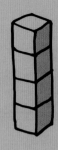

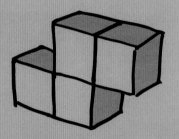

- Establish a culture that is conducive to and acknowledges the courage to draw.
- Create simple and clear drawings. Visualization is not art but a means to an end.
- Systematically replace key concepts or sentences with images, icons, and stick figures.
- Take advantage of the impact of images on our brain: Images leave a deeper impact.
- Solve communication problems on interdisciplinary teams with drawings; this way, you can overcome language and cultural barriers.
- Use drawings and visualized stories to communicate knowledge in meetings.
- Use simple drawings and graphics.
- Because it's faster: Sketch something instead of cumbersomely writing it down.
- Visualization is like a language: You have to practice and apply it. Just do it! Now.

Good stories have accompanied humanity for thousands of years. In olden times, the profession of storyteller used to exist. Today, books and new digital media have replaced this profession. But we still like good stories. Pretty late it was discovered—and accordingly used— that products and objects can tell great stories as well. Architects such as Gaetano Pesce, one of the icons of Pop Art design, once remarked that we are separated from objects as long as consuming them is the only primary reason for their existence. Why do we take Pesce as an example? Peter has told us about his enthusiastic admiration of Pesce. He was particularly taken with the "La Mamma" chair, UP 5, from 1960. We'll leave the question open whether the chair has any resemblance to Priya at this point; it's up to your imagination. We will discuss later why imagination is important and what all this has to do with storytelling. First, some information on the UP 5 chair: It has the shape of a woman—hence the name "La Mamma"—and was inspired by prehistoric fertility goddesses. To implement his idea, Pesce used an innovative technology that allows the creation of large foam-molded parts without a load-bearing inner structure. Thanks to a vacuum chamber, the piece of furniture could be reduced to 10% of its volume and thus shrink-wrapped in air-tight film. This made it easy for the buyer to get the piece of furniture home. Only once the film was removed did "La Mamma" unfold to its original and final size and shape.

LA MAMMA , PESCE 1969

Until the 1960s, design was defined by the actual function. Many designers perceived the relationship between the consumer and their products primarily through their actual use. It was only in the 1970s that some designers challenged this paradigm. They added "nonfunctional features" and artistic elements and ornaments to their objects. This showed that the relationship to an object can consist of more than merely its primary functionality. Since then, functionality often has not taken center stage. A more holistic relationship with the object emerged, with a deeper meaning for the consumer. Without this deeper meaning, a product is not perfected in the eyes of the consumer. This means that the consumers became the actual designers. They create the meaning of an object through the intimate relationship they establish with it. Lilly has observed this behavior often in her younger students, who have such a close relationship with their smartphones that they give them pet names.

CARLTON, SOTTSASS 1981

When does imagination influence the buying process, and at what point is storytelling used?

A headline in *Forbes* magazine claimed that good storytelling can heighten customer acquisition by up to 400%. Now we all see the $ sign flicker in front of our eyes, which ought to make us sit up and take notice. With all services and products, the ultimate art is to maintain the desire for them, which is based mostly on the relationship between the object and the consumers. Desire is exactly the condition that the object or product does not have, though. Instead, it is the presumed relationship with the object that satisfies desire. With his dream car, an American electric vehicle, Peter is one of these modern consumers who "cultivate" their desires. Any product with hitherto unknown features has the potential to fill us with enthusiasm beyond the limits of reality. We can compare this state with daydreaming or the merging of feelings of happiness from fantasy and reality. In general, we are all confronted with a dilemma that becomes manifest in the desire to own the object and the fact we don't own it.

What happens after we have bought the product? Why do we want to own the product at all? Haven't we already experienced a feeling of happiness with the product in our imagination?

It seems we find the perfect experience more with new products than with goods we already own, which have lost the capacity to embody the perfect experience. With the actual use of a product, we have the opportunity to experience our fantasies and dreams, which we have built up beforehand and that revolve around the product. This doesn't mean that a product will become interchangeable as soon as we have bought and used it. Consumers have the power to turn their personal objects into something special.

How can stories of products be put in the right context?

Stories are an excellent means of describing the various relationships between consumers and products. For some products, especially when it comes to fashion, a good story is of greater value than the function or the quality of the clothes. With the help of stories, consumers can identify with the clothes they bought and display their fashion style to the outside world. Storytelling has the potential to speak to the audience like this: **"Hey, it's really only about you!"**

In general, we distinguish **three types of stories**, which are important in the perception of products:

1. **Commercial stories from manufacturers** such as Coca-Cola, who use them with great skill and cleverness in their marketing. We all know the famous taglines of Coca-Cola such as "Release the brrr inside you" or the Coca-Cola Christmas stories, which for generations have become an integral part of the run-up to Christmas. Although Coca-Cola today has become more personal—through promotions such as "Share your family photo," "Your own name on the bottle," or the current series of "Drinking a Coke with your friends"—we have stored these images deep in our unconscious.

2. **Lifestyle stories from and about users**. These stories are often associated with emotional goods such as cars, motorcycles, watches, and other luxury items. Consumers buy these goods to pursue a specific lifestyle. The Chevrolet commercials "Maddie" and "Romance" are good examples of this because they tell profound and unforgettable stories. The stories build a close relationship with the consumer far beyond the actual product. Often, the stories are supported by so-called fan clubs.

3. **Stories with the character of a specific memory**. These individual stories are based on the personal memories of things past and vary from individual to individual.

EXPERT TIP
How can empathy be understood
as a design paradigm?

A good story or a story that is told perfectly always follows a typical narrative. In so doing, an arc of suspense is created, so listeners remain attentive.

This arc of suspense is essential and is built up continuously from the first moment—right up to the final punch line.

A good story that works usually includes five elements:

• an emotionally significant initial situation;
• a (likable) main character;
• conflicts and hindrances, which the main character must overcome;
• a recognizable development and change ("before and after" effect); and
• a climax, including the conclusion or the moral of the story.

Good stories induce emotions in the viewer and convey a message. To be able to tell a good story, we must know our target group quite well. Again, it is of great importance to have built up empathy beforehand. The topic of empathy has already been extensively described, so in this section we will put empathic design in the context of other design approaches.

Many approaches have contributed to the development of empathic design or are based on similar thoughts.

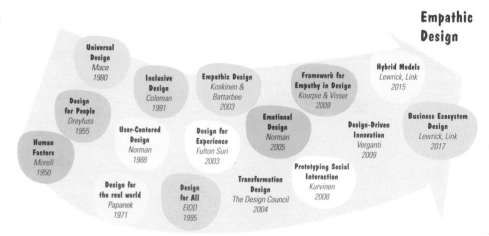

Empathic Design

Empathic design is the development of products and services that are based on unspoken customer needs. New tools have been added over the last few years that allow companies to understand the mood of the customer, making it possible to experience a situation from the customer's point of view. This experience frequently yields important product information, which cannot be tapped by means of normal market analyses and well-known empathy tools.

In many companies, such approaches have become an integral part of product development. The use of so-called third-age suits is a good example. They allow designers and product managers to experience with their own body the limited physical capabilities of seniors. There are also methods that require less technology and focus on specific senses. The goal of geriatric sensitivity training is to make certain physical states palpable. Glasses that simulate the clouding of the cornea or age-related macular degeneration can help users experience how these impairments affect everyday life. Alongside the glasses, there are gloves that simulate limited sensitivity and headphones that replicate a hearing impairment. The experiences are conducive for the development of products, services, and processes.

For all of us, it is easier to become motivated when we can imagine the purpose of our actions. This way, the belief that we can achieve a set goal is strengthened. For this reason, it is always advisable to start with the Why. In the **model of the Golden Circle**, the Why is in the center. The limbic system (**Why**) is located in the center of the brain and is guided by emotions and images. We're dealing with behavior and trust, emotions, and decisions here.

Successful companies keep the Why on center stage with a clear vision. In such companies, employees know why they get up in the morning and go to work. Spotify, for example, has the great mission of bringing music to the world.

The **How** describes how the work is done and which particulars it involves result from the Why. Apple is another much-cited example in this respect.

The logically thinking brain (**What**) is located on the outside in the model of the Golden Circle. It encompasses rationality, logic, and language. The How connects the two elements and explains the process of how something is done.

The inventor of the Golden Circle, Simon Sinek, expresses it this way: People don't buy what we produce—they buy *why* we produce something. This is why we should always start with the Why. The same model applies to internal communication (e.g., to digital transformation). Successful business leaders communicate from the inside to the outside in the model of the Golden Circle. Employees know why they do something, how they do it, and what they do.

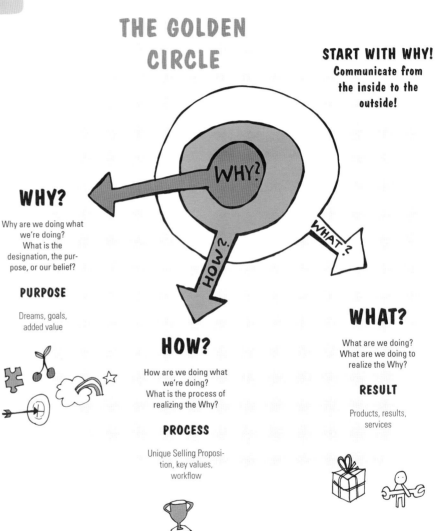

THE GOLDEN CIRCLE

START WITH WHY!
Communicate from the inside to the outside!

WHY?

Why are we doing what we're doing? What is the designation, the purpose, or our belief?

PURPOSE

Dreams, goals, added value

HOW?

How are we doing what we're doing? What is the process of realizing the Why?

PROCESS

Unique Selling Proposition, key values, workflow

WHAT?

What are we doing? What are we doing to realize the Why?

RESULT

Products, results, services

EXPERT TIP
How can empathy be understood as a design paradigm?

A tried-and-tested tool for the generation of emotional stories is the Minsky suitcase.

Do we know where our suitcase is at the moment?

Most of us are likely not thinking about our suitcase just now. It is somewhere in the basement or stored in a closet. Nothing unusual here. Once life has returned to the daily routine after vacation, the memories of fine dining on the French Riviera or the white sand beaches in the Maldives quickly start to fade away. The last memory of the vacation consists of a few grains of sand hidden in the inner pockets of the suitcase. For a certain period of time, our suitcase was a synonym for a different lifestyle, a better life, life as it ought to be: with pleasure, relaxation, an uncomplicated and free schedule.

Maybe we never actually thought about it, but the items in a suitcase have basically four different types of use:

1. things of everyday life (toothbrush, socks, change of clothes),
2. things that are very important to us and do not take much space (a photograph, a lucky charm, or a diary),
3. things we want to impress people with (jewelry, a fashionable scarf, cool sunglasses), and
4. some free space for things we want to purchase on our trip.

A packed suitcase is the compressed version of our personality:
It is orderly, chaotic, an imitation, an original, it bears traces of past adventures, and so on. When we travel, each of us has exactly the suitcase that fits us best and is thus a mirror image of our life.

To tell emotional stories, a suitcase can be an inspiring starting point. We find old suitcases in the attic or buy a new one at the flea market. In a second step, we build up a relationship with the suitcase and its contents. Why was the suitcase forgotten under the roof? What could its story be? We take some time and write a small fictional story about the object and its possible relationship with its former owner.

Let's assume there's an old, heavy winter coat in the suitcase, and our design challenge is to create a new soap. No restrictions are imposed on our design team in terms of the shape, smell, color—but it's not only the soap that is to be designed. The packaging and the marketing concept must also be created. The following story might have emerged from the inspiring framework of the old winter coat:

"An old lady is looking outside the window during the winter. She sees the road covered in ice while she is preparing for a dinner with her grandchildren. She is really looking forward to seeing them again and would like to be the best host she can be. However, she suddenly realizes that she forgot an important ingredient. She feels stressed because she will have to go out in the cold winter weather to buy the ingredient…"

Two examples of a possible product are the "Savon 1890," a very simple, old-fashioned, handcrafted soap in plain packaging, and "Soap Crystals," which are based on the experience with an old walking stick.

We all know the problem of how the respondents in a direct user survey describe their own behavior as an ideal type but do not show their true self. We ask them about their goals and desires, but the reply only consists of the most obvious insights. One way to reach users on a more emotional level is to offer stories of their dreams, which give us the opportunity to learn things that go deeper and reveal their true needs and desires.

A project with the name "wearable dreams" is a good example of how such stories about dreams can provide an inspiring framework for a design thinking project. In this project, the interviewees were initially asked to imagine that their favorite piece of clothing was a person. Then they were asked to describe the personality of this person:

- What is the name of the favorite piece of clothing?
- How old is it, and what does it do for a living?
- Is it pretty shy or rather extroverted?
- Where was it born, and what is its marital status?

This way of talking about a product helped the interviewee to think about his favorite item of clothing and thus transported the object into a social and emotional context. The remaining interview built on these dreams. The respondents were asked to imagine the person depicted by the item of clothing in a difficult situation. Fortunately, the person had super powers that got him out of the situation.
Initially, the respondent was asked to describe a situation that he or she did not want to get into. In addition, the person was requested to choose a specific role and take it on. The questions were, for instance:

- How do the surroundings look?
- Are there any other people?
- What items are lying about?

In the ideal case, the respondents wrote down a little story they made up about how the person got out of the difficult situation. They were asked not to reflect too critically on what is possible or impossible in reality. Ideally, the whole thing was pepped up with drawings. The length, content, and depth of the story were irrelevant.

The design process builds on this information. The idea behind it is that objects must satisfy our emotional needs, for one. Also, our rational stories are the best way to transport these needs.

My favorite piece of clothing is a bucket hat. Its name is Alex. It is 34 years old and a small-time crook!

THUG LIFE

Design trends such as a popular style, and hip combinations or colors are not the real trend. These attributes are only the tip of the iceberg. To identify the true trends, you have to dig deeper. This is the only way to reveal the artifacts. Changed behaviors, beliefs, and social forces make up a trend.

We know scenarios as descriptions of alternative possibilities, based on which decisions for tomorrow are made today. They aren't forecasts or strategies but are more like hypotheses about various maps of the future. They are described in such a way that we are able to identify the risks and opportunities in terms of certain strategic realities. If we want to use the scenarios as an effective planning tool, we should design them in the form of captivating and, at the same time, convincing stories. These stories describe, for instance, a range of alternative future scenarios that will lead the organization to success. Well-thought-out and credible descriptions help the decision makers immerse themselves in the scenarios and perhaps even acquire a new understanding of how their organization can master possible changes on the basis of this experience. The more decision makers we introduce to the scenarios, the better they recognize their importance. Moreover, scenarios with easily comprehensible contents can be taken quickly into the organization as a whole. These messages stick easier in the memory of employees and managers at all levels.

The use of future scenarios for visionary projects differs from the daily work in project or product management. The scenarios constitute an inspiring guide into a possible future. Visionary projects not only serve to inspire the entire organization and challenge existing technologies; they also help to galvanize individual employees. Thus the future scenarios seem to have a great influence on the organization; however, they are more difficult to orchestrate because they deal with the unknown. Organizations are often unable to initiate the transformation and fall right back into their daily routines—not least because the future changes had not been sufficiently prepared for. To avoid this relapse, companies like Siemens publish "Pictures of the Futures" at regular intervals.

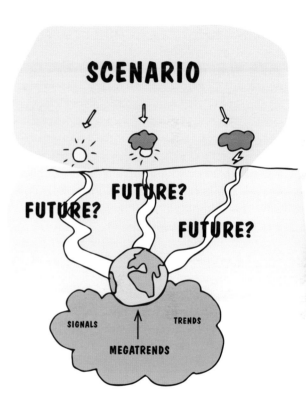

Pictures of the Future (Siemens) link realistic current trends with distant future scenarios to align and direct business activities. For one, the future scenarios created can be used well to formulate or redefine the starting question in design thinking, and can give further momentum to the process of creative problem solving on the team.

Step 1: We extrapolate from the world of today

We start with the daily business of our company and look at the trends, from which we extrapolate how the near future of our company might look. Data and information from different sources, such as industry reports and interviews with experts, are analyzed. The fastest way to reach our goal is to fall back upon the known trends in an industry, like internal trend reports and market analyses, which are freely available on the Internet. For example, we take the general Gartner Technology Hype Cycle as a starting point. Our best course is first to compile a provisional list of trends; discuss them briefly on the team; and note the estimated importance, strength of impact, and the degree of maturity of the relevant industry.

Step 2: We apply strategic visioning

We detach ourselves completely from our own business focus and our own professional blindness and design various distant future scenarios with a true outside-in perspective, independent of our own company (in the example, four scenarios have turned out to be ideal). Because we are dealing with distant scenarios, elaborate studies are usually carried out with the inclusion of worldwide research. Luckily, Siemens has already done this work for many industries with their Pictures of the Future and has made the results available free of charge (among others, from the sectors of energy, digitization, industry, automation, mobility, health, finance, etc.). We choose a positive, constructive, and profitable scenario and ask ourselves: "How might our company make a maximum contribution to this scenario? What would we have to do and offer?" We stay in the future in our thoughts and do not allow the processes and structures of our company today to influence us.

Step 3: We "retropolate" from the world of tomorrow

We carry out a retropolation from the scenarios. The point here is to draw conclusions for the present from the "known" facts of the future scenario. We juxtapose the results from step 1 with those of step 2, combine them, and infer from that what it means, in very specific terms, for the alignment and direction of our company today. In which directions should we innovate and do research? What skills must be developed? What personnel should be hired? And how should processes be redesigned so we are prepared for coming challenges and opportunities?

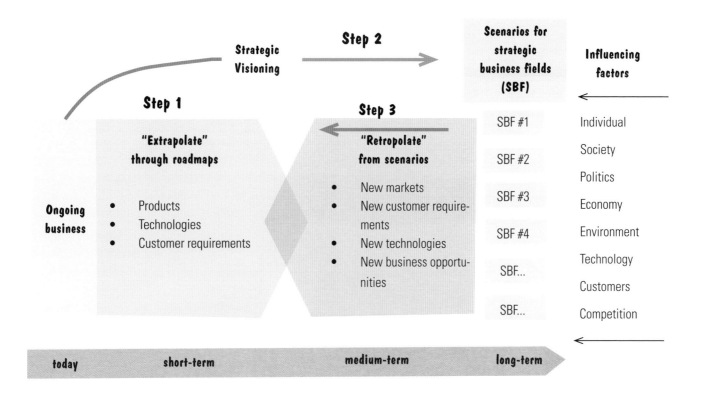

Step 2

Strategic
Visioning

Step 1

Step 3

Scenarios for
strategic
business fields
(SBF)

Influencing
factors

**"Extrapolate"
through roadmaps**

**"Retropolate"
from scenarios**

SBF #1

Individual

SBF #2

Society

Politics

SBF #3

Economy

SBF #4

Environment

SBF...

Technology

Customers

SBF...

Competition

Ongoing
business

- Products
- Technologies
- Customer requirements

- New markets
- New customer require-
  ments
- New technologies
- New business opportu-
  nities

today   short-term   medium-term   long-term

177

Well-thought-out digital storytelling is becoming more and more important. After all, we use different digital tools every day and consume a correspondingly high amount of digital words. Digital storytelling gives us the opportunity to represent our company perspectives in more detail and use emotions in order to get more attention.

Storytelling consists of two words: "story" and "telling"—content and performance. We know traditional storytelling with a narrator, who performs in front of his audience. Nonverbal reactions help the narrator to assess how well the listeners are following him, so he can react spontaneously. The digital world has none of these nonverbal reactions. We must use other tools to establish empathy with a digital potential audience.

There is a broad range of media we can use, from multimedia films to audio broadcasts all the way to webinars. To select the right content and media, it is important to develop a deep understanding of the target group. We recommend creating so-called buyer personas and getting information from potential customers:

- Why buy from us?
- How do customers find us?
- What questions are we asked during the sales process?
- What motivated the customers to search for a solution?

Because all of us are addressed on multiple levels, emotional and intellectual elements associated with our brand are equally important. It helps in this context to flesh out the storytelling with data and facts. We also have the option of encouraging our users to generate content.

Lego provides an intriguing example of a digital story:

Problem:	Lend a new profile to an old children's toy
Campaign:	90-minute "Lego film"
Agency:	Warner Bros., Hollywood, CA
Solution:	A good movie for young and old with the message that we are imaginative builders at any age.

**Mindset
Digital storytelling:**

1
Keep your story or message short!

2
Make sure the content of your story is linear and the narrative is clear.

3
Show, don't tell!
Use images to give the story more content!

KEY LEARNINGS
Tell stories

- Don't just focus on the form and material but turn the product into an experience. The goal is to stimulate the imagination of the consumers.
- Speak to the various senses in order to create a holistic experience for the user.
- Make use of factors of success such as focus, simplicity, interactivity, and branding for a good story.
- Create future scenarios for an inspiring framework. They help to enshrine and communicate a vision.
- Establish empathy with the user. Empathy is the basis for every story because consumers want to satisfy their needs. Arouse fantasies and desires.
- Try to tell a vivid and exciting story by pepping it up with other people surrounding the user.
- Get inspiration from tools such as the Minsky suitcase. It helps in getting new insights and thus creating stories.
- Thousands of digital words are consumed every day. Turn digital stories into a pivotal medium in order to heighten user attention.

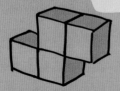

2.5 How to trigger change as a facilitator

We all assume the role of facilitator sometimes. For example, in this function, Jonny has invited Lilly to a design thinking workshop for developing a creative space. Marc has found a suitable team for the start-up at d.school this way. Thus the work of a facilitator is very au courant at many companies and entrepreneurship programs. The trend can also be seen from the fact that Peter is currently getting a truckload of offers for further training in this context. The offers range from courses on Theory U to the art of hosting camps. With respect to the latter, the facilitator is also referred to as the host, who ensures that all those involved feel good and safe in terms of the change. These new concepts sound rather esoteric to Peter's ears but he once thought the same of design thinking, and now he champions this mindset with total conviction.

Is there is a typical facilitator attitude that is vital and important for change and transformation?

Peter is aware that, as a co-creation manager and in his role of facilitator, he triggers the "ignition" of new ideas. He makes it possible through dialog, clarity, involvement in the problem statement, and promotion of active participation. He supports the team in channeling a wide variety of opinions, which, in the end, leads to outstanding solutions.

Facilitation results in more sustainable decisions, which are backed by many. This means the highest added-ed value a facilitator gives consists of creating the necessary structure and culture of dialog, so the team can focus on finding the best decision for their problem statement.

Discussions and exchange of ideas can be subdivided into two categories. In the first, there are those in which a decision takes center stage. Discussions that focus on the exchange of ideas and information are different and fall under the second category.

The implementation of changes succeeds when the employees are involved in it in a coherent and consistent way. The key to a company's success does not only rest in new products and services, but also in how organizations integrate the intellectual capital of their stakeholders in change processes.

This is why a facilitative attitude and the corresponding methods and approaches are seen today as a critical factor for the success of organizations and companies.

Each of us has different ideas of how a decision-making process would look in an ideal world. Some of us have the notion that decisions should be made through a logical chain of ideas, opinions, and analyses. According to this idea, all of us in a group think at the same speed, move forward linearly, and begin with a question at the same time in order to get to a solution at the same time.

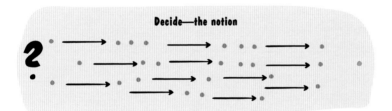

Another mental attitude follows the principle of hope. This principle consists of the idea that group members have different opinions, which nevertheless can be brought to a common denominator. A solution is found without much divergence and with little effort.

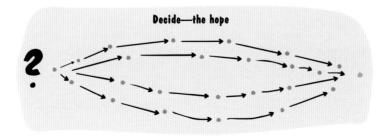

In many situations, we are confronted with huge problems for which there are no easy solutions. These issues are complex. They require a great deal of endurance, and their solutions are based on a multitude of ideas and opinions. Peter has often gone through this situation when presenting decision memos to top management for the solution of a wicked problem, such as reducing the traffic in large cities through extensive digitization solutions or new technologies. Often the response is a killer phrase, like "That will never work!" or "The market is too small with too many stakeholders." This happens whenever decision makers are incapable of mentally penetrating the solution or are unwilling to do so, or if they are driven by a fear that the changes might go deeper than desired. Complex interrelationships are strenuous and often difficult to understand!

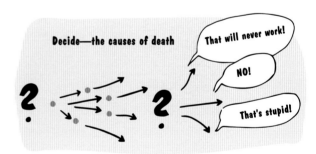

In management meetings, Peter time and again observes that, although an issue is discussed, the decision is postponed indefinitely in light of a rather difficult basic emotional mood. Or a decision is made that wasn't even discussed. Often, the boss makes a decision even before the multitude of different ideas from the divergent phase has even been thought through.

What's so problematic about this is that all the divergent energies and ideas slow the entire process down. The divergent ideas that have not been discussed get dug out again in every project phase and brought into the process.

With such decisions, the team is often still busy generating ideas—the supposed solution was actually quite far away. In Chapter 1.2, "Why is process awareness key?" we already talked about the groan zone. Once again we must emphasize that it isn't easy for teams and groups to accept and engage with new and contradictory ideas. You want to get a project moving, but you notice that the team's energies go in all directions and are not to the point.

At the moment of ideation, usually group members do not have any clue where they are headed. Especially with complex questions, this situation is perceived as unpleasant, difficult, and simply awful. Groups often experience this situation as dysfunctional. But it isn't that way at all. Every group and every team go through this period of emergence. The facilitator helps everyone endure any irritation, confusion, and disruptions.

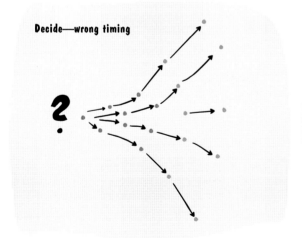

Decide—wrong timing

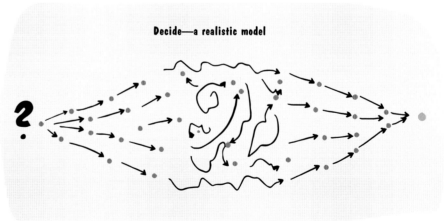

Decide—a realistic model

The nine principles of facilitation

Facilitators use different methods and approaches, which are based on nine principles. These rules are seen as the golden thread for the facilitator:

1. Assumptions and conclusions

We continually make assumptions, use attributes, draw conclusions, or are subject to well-known prejudices. That is not the problem. What makes things difficult is we're not aware of it or believe our assumptions are the truth. In effective groups, these assumptions are reviewed and tested again and again.

2. Sharing of relevant information

This concerns not only data and information directly related to the question, but also all information that could influence a process.

3. Use of specific examples

In many projects, information and data crop up in a nonspecific manner, excluding information such as background, author of the information, location of the action, and much more.

4. Explanation of the intention and the conclusions

Our intentions indicate the purpose we pursue. When we explain intent and conclusions, we share our insights with the group on how we got to a certain piece of information and how we drew our conclusions. This way, groups become more open to different perspectives.

5. Focus on interests, not on positions

Interests have something to do with our needs and desires. Thus we refer to the relationship we have with a given situation, while positions must be seen as adamant opinions about a situation. Effective groups convey their interests in order to develop common interests.

6. Combination of advocacy and inclusion

In groups, contributions to the discussion often turn into a series of monologues, instead of ending in a true exchange. To advocate something while creating a reference to the other contributions allows for effective and common learning and understanding the issue on the next higher level.

7. Finding a design for the next steps and testing the differences

Groups decide by themselves which core themes should be discussed, when and how this will be done, and in which way different perspectives can be put next to one another without hindering the cognitive process.

8. Discussion of topics that can't be discussed

Groups always have core themes that bedevil them and that they are apparently unable to discuss because they fear losing effectiveness. Groups can be empowered to confront even topics that seem completely impossible.

9. Support of decision-making processes on the basis of an adequate commitment level

We know different routes and types of decision-making processes (e.g., delegation, consensus, democracy, consultation, advisory process). The degree of acceptance ranges from resistance to noncompliance to compliance all the way to internal commitment.

Facilitation can be useful for all types of transformation and change and all questions entailed in them in companies and organizations—from the development of a corporate culture all the way to strategy definition.

How do we put together the team to promote dialog?

The process always takes center stage for the facilitator. In terms of content, the facilitator stays neutral. Facilitation always assumes that expertise, knowledge, and in-depth insights are available in the company itself. A facilitator creates the space for us to be able to exchange ideas in an adequate system. This exchange has the goal of enabling consistent, precise, effective, and successful collaboration.

According to the ARE IN formula, an adequate system consists of participants representing the following:

A uthority—who has the power to initiate change?
R esources—who contributes specific necessary resources?
E xpertise—who has experience and a very extensive range of knowledge?
I nformation—who provides us with information, including the informal kind?
N eed—who knows the needs of our customers and users?

The facilitator's task is to make better use of the resources and potentials existing on a team or in a company. Hence he will direct developments based on strengths and not based on the prevention of weaknesses. The principle of facilitation is thus resource-oriented, not deficit-oriented. Facilitation is the opposite of almost all well-known consultancy approaches, all of which are more or less based on deficit orientation. Practically every consultancy is for compensating for deficits in the company, not for tapping resources that are already available.

In addition, facilitation is based on some very specific assumptions about the company and about the nature of changes:

- Trust the process.
- The knowledge of change rests in the system.
- Have a low profile as a facilitator and don't push yourself to the front.
- Build up a community before decisions are made.
- Control what you can control, otherwise let it be.
- If a method or intervention does not help the group, forget it.
- What we focus our attention on will become reality.
- People want to take responsibility and do something meaningful.
- Everybody is doing his or her best.

Based on these assumptions, facilitators develop specific approaches to initiate participation processes and to support teams.

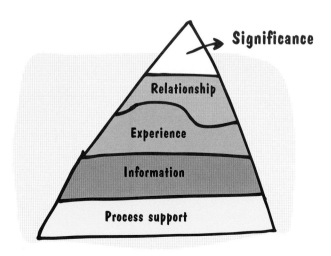

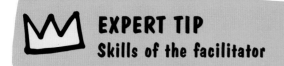

EXPERT TIP
Skills of the facilitator

What skills must a good facilitator have?

Facilitation is based on six fundamental skills:

1. Create relationships

This involves the building up of constructive collaboration: the development of a consensus on the purpose, goals, roles, and responsibility—in other words, making clear what values are important for the collaboration.

2. Suitable processes and methods

The planning of suitable group processes and the selection of the right methods allows for open participation. It is of fundamental importance here to understand how those concerned can be integrated in the process in order to support different styles of learning and thinking.

3. Participation-oriented environment

In varied participatory processes, the exchange and the collaboration between all those involved is promoted. This includes the use of effective communicative skills as well as the feedback regarding commitment and behavior. Diversity is rewarded, and conflicts are monitored actively.

4. Meaningful results

Meaningful results emerge through the use of suitable methods and adapted process steps. It can be helpful here to lead the team or group back to the original question. Self-awareness of the group can also be very conducive to reflection on one's own experiments and solutions.

5. Breadth of knowledge

Facilitators can take advantage of a great fountain of knowledge in terms of facilitative methods; they know solution processes and decision-making processes; and they are experts in differentiating between process, task, and content. They work out new processes, methods, and models to meet needs even better and practice continuous reflection and constant learning.

6. Positive attitude

Facilitation refers to a positive attitude that is exemplified by the facilitator's own behavior, such as a high level of congruence between actions and personal values, as well as the ability to reflect on the needs of the group. The facilitator notices at the right point in time whether he has been bypassing the needs of the group or not.

HOW MIGHT WE...
prepare a workshop optimally as a facilitator?

What must be made clear before starting a workshop?

Before we select a specific workshop or a specific intervention or method of facilitation, we ought to find out exactly what is to be achieved, how it is to be achieved, and why it is to be achieved. The more clear-cut the specifications, the more successful the implementation.

Prior to a moderation in front of large groups, we should collect information on why such a large group of people is necessary. The purpose must be attractive and meaningful for all parties involved. It should not be formulated and structured within limits that are too narrow and must contain potential for exploring and discovery. If everything remains the same afterward, the facilitation has failed in its purpose. The following keywords help to clarify the purpose of an intervention. Is it about

- developing awareness?
- finding problem solutions?
- fostering the development of relationships?
- initiating the exchange of knowledge?
- supporting innovation?
- developing a vision and sharing it?
- clarifying the development of capabilities?
- building up the development of leadership?
- solving conflicts?
- drawing up and enacting strategies or actions?
- expediting decision making?

In general, the following success criteria must be borne in mind:

- There is a high degree of mutual exchange.
- The establishment and the deepening of relationships is elementary.
- Everybody sees themselves as learners and contributors.
- Everybody is involved (in the discussion, in drawing, listening, talking).
- Everybody will be heard.
- Different perspectives are perceived as such.
- Shared findings will be condensed.
- Everybody is clear about what comes after the workshop.

How can we proceed in the workshop, and what questions must be answered?

The questions of How, What, When, Why, and Whom are focused on:

- What is the change about?
- On which playing field will we play?
- What changes are possible, and how should the goals look?
- What does success mean?
- How will the company look after the process?
- What is today?
- What does the reality look like?
- What strengths can be identified? What weaknesses?
- What are the benefits to be developed from the process?
- What needs are to be addressed?
- Who benefits from the results? What are the risks of the project?
- What was good? What can we do differently next time?
- What will we do next? How do we want to proceed?

What methods can we use as facilitators?

There are countless approaches, methods, and associated variants that can be used. In the end, we have our own toolbox of methods, and must use them in a targeted way.

One possibility is visual facilitating and graphic recording. As the name suggests, the issue is to visualize information and dialog in real time and directly on site. Visualization mainly serves the purpose of transforming complexity into a structured picture. We particularly recommend this method for large projects with a great need for change management and in a difficult dialog that focuses on documenting the decisive turning point.

We all know the phenomenon of the most optimization potential being hidden in processes that we believe are running perfectly smooth. Still, we hesitate to change systems that seem to be working.

A way of thinking and acting that can help us break from our habits is appreciative inquiry. We focus on examining the given facts; that is, all those things that work perfectly in a system.

As participants in workshops, we have all taken part in world cafés, open spaces, or art of hosting events. What all these concepts have in common is the idea of seeking dialog in a circle (the circle way). The advantage of a circular arrangement: Participants collaborate more actively and are more inclined to take on a leading role with regard to a topic or argument. We recommend such a setting for all group work and, in particular, in change-intensive projects of organizational development.

One method that allows us to proceed in a way more driven by impulse than linearly is dynamic facilitation. This approach consciously appreciates mental leaps on the part of the participants, which are noted down ad hoc in lists.

Usually, documentation is done in four lists with the thematic blocks of:

1. Questions and challenges
2. Initial ideas and solutions
3. Concerns and objections
4. Information and points of view

The challenge for us as facilitators is that we pick up information quickly in such a setting and are forced to give room to reflection constantly.

Frequently, we have observed the effect that the best dialogs occur during coffee breaks or with an aperitif after a long and rather tedious workshop. The idea of open space technology as a format picks up on this idea and designs the workshop per se as a free space in which we find solutions in common. We recommend an open space approach for the final presentations of design thinking projects, for instance. The visitors can explore the ideas themselves, and their curiosity lets them approach the topics. For a structured sequence, it is advisable to organize a **world café**. The objective is to share knowledge among different groups of people. After the discussion round, the participants are moving from one table to the other, while the table host facilitates the conversation with the next group. The idea behind it is the same. Again, the issue is to speed up an informal and nonetheless intense discussion in the small circle.

WELCOME TO THE
WORLD CAFÉ!

What do we have to pay attention to in the individual phases?

We have all organized workshops at one point and assumed the role of moderator or facilitator. Basically, there are four simple steps in the planning and implementation that we would like to summarize again as follows:

1. **Determine the context**
2. **Carry out the planning**
3. **Implement according to demand**
4. **Initiate reflection and learning**

In the end, the important thing is that we have achieved the right results and created a "Wow!" experience that either generates a momentum for changes or else has taught us something upon which we can build.

The facilitator always concentrates on the group and tries to encourage, empower, and enable it. These are the three Es of the facilitator.

ENABLE!
EMPOWER!
ENCOURAGE!

1. CONTEXT	2. PLANNING	3. IMPLEMENTATION	4. REFLECTION
• Why? • What are we supposed to achieve?	• What exactly? How? Who? When? Where? • Who does what? • What do we need?	• How is the group doing? • Are we in flow? • Will we achieve the goals this way? • Do we have to adapt?	• Goals achieved? • Next steps? • What went well? • What would be even better?
• **Assumptions** • **Desired result** • **Goals**	• **The right process** • **The right participants** • **The right environment/ space** • **The right information**	• **Preparation, e.g., space** • **Welcome/warmup** • **Sequence/methods/ facilitation** • **Conclusion**	• **Reflect and learn** • **Continuation**

KEY LEARNINGS
Perform facilitation

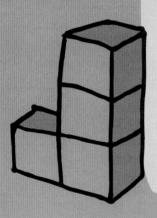

- Create relationships and foster the participation of each and every member.
- Clarify the meaning and purpose prior to the workshop.
- Plan carefully the process, the selection of participants, the environment, and the necessary information.
- Pay heed to a good team makeup. Use the ARE IN formula.
- Use tools and methods such as visual facilitating or world café, adapted to the situation.
- Create space for diversity: for example, of cultures, points of view, genders, nationalities, hierarchy levels, and functions.
- Use creativity methods, such as brainstorming, and guide the participants through difficult phases in the process (e.g., the groan zone).
- Assume a positive attitude and ensure the well-being of the participants.
- Always keep the nine principles for facilitators in mind (e.g., that relevant information is shared and that the goals of the workshop are clear to one and all).

By now, Peter has carried out many projects with design thinking, and innovative and customer-centric solutions have emerged. His environment, his direct supervisors, and colleagues also know that design thinking is an asset for the company. However, Peter notices more and more often that not all teams back the mindset of design thinking.

In discussions with like-minded people, at conferences, or in forums, he realizes that design thinking can bring about relevant solutions; yet many organizations have a hard time disseminating the approach transversally. Resentment grows, and solutions are sought to change the mindset.

At a recent meeting of the DTP community in Zurich, a colleague of Peter's who was an enthusiastic cyclist came up with an apt metaphor:

"Design thinking is like a fantastic racing bike and it effectively takes us where we haven't been before! But the fact we own a racing bike does not mean we'll be able to cross an Alpine pass. We must have the corresponding fitness for it!"

Peter is convinced that his organization—like many other companies—does not have the necessary fitness to live design thinking all the way, with all its consequences.
Upon closer inspection, Peter quickly realizes that his organization has too many departments that do not share the same mindset: Although they have a great racing bike, they have a beer belly and no fitness whatsoever.

So what is the problem that prevents design thinking from spreading?

The problem Peter shares with others who are responsible for innovation is the form of the organization in which they work. The company has a typical silo structure, which has become manifest over the years and allows management to deal with growing complexity and the requirement for more efficiency. It is made up of specialized teams that optimize their own fitness. The preferred tools are process improvement and operational excellence, which make the silos even more efficient. Transversal collaboration for creating a consistent customer experience falls by the wayside.

To overcome such a silo mentality, we must undertake consciously designed measures and initiate change processes that make cross-departmental collaboration possible. This is the only way to establish a new mindset in an organization as a whole. Design thinking is always only as effective as the capacity of the organization to implement the result comprehensively and evenly.

SWEATY

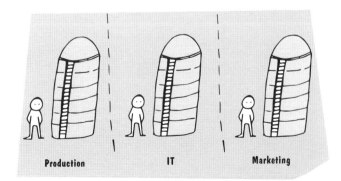

Production   IT   Marketing

How can the change be tackled, and why are companies so often held hostage to this problem?

Many companies have a diversified organizational form in which the individual business units differ greatly and cultivate their own work processes and subcultures. Although this form of organization can help to guide a growing organization into manageable channels, the separation of the organizational elements can result in the fact that the overriding meaning of the company gets lost on the way. The business units and departments see the objectives for their work exclusively in relation to themselves. Objectives that are meaningful to all organizational units are not universally lived. If they do exist, usually they're only the key financial figures of the company—such as profit and EBIT—which are set for everybody in the company as orientation and corporate objectives.

The consequence is that an integrated and coordinated way of working across the entire organization is quite difficult. Moreover, human, nonfunctional relationships fall by the wayside due to the lack of motivating goals.

How do we react to changes in value creation?

The customer experience is becoming the primary product in many segments owing to the transition from industrial manufacturing toward servitization (alignment to goods and services).

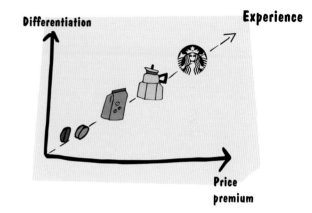

Economic success is determined not by the quality of the product but by the fulfillment of our needs across the entire customer experience chain. Customers want an experience—no matter what kind—that they can share with others and that enables them to fulfill their wishes. For this reason, customer centricity is (again) becoming one of the core issues of management in the experience economy. The mindset of design thinking can make a crucial contribution to the development of customer-centric solutions.

What is it that successful businesses do differently?

Successful businesses gear all their activities and all areas as well as their employees toward the customer. In addition, they integrate the customer and his needs deep in their strategy, such as by creating an enhanced awareness of strategic foresight. They put the focus on customer interaction and the design of worlds of experience.

In many companies, this requires a radical change in the usual understanding of leadership, away from the leadership doctrine of an omnipotent management toward a leadership culture (and a mindset) that enables the organization to overcome this functionally acting structure. The change in the understanding of leadership is also seen as a necessary step toward an integrated organizational form in which employees develop a high degree of intrinsic motivation but, at the same time, direct their activities toward an overriding goal shared by the entire organization.

A unifying and integrating meaning that can be enshrined using the mindset of design thinking helps to implement transformation. As a creative tool, design thinking can fulfill a methodological role in the transformation of the company by providing important tools. The human-centered approach helps to establish orientation toward the customer, which includes considering colleagues from other departments as customers as well. In our experience, effective design thinking can optimally unfold only in an integrated organization.

The path to the integration phase, however, starts in the so-called pioneer phase, in which the organization is often built around a leading figure. Then the company grows and diversifies into different areas. In the process, different cultures and silos evolve. This phase is characterized by efficiency and effectiveness. Only then can an organism come into being that ensures that these systems are perfectly matched with one another. For this reason, organizations must be reorganized and rebuilt at regular intervals. In nature, an apple tree must be cut and pruned for it to yield a rich harvest again and again.

1. PIONEER PHASE

An organization built around a leading figure

>> family

2. DIVERSIFICATION PHASE

Emergence of a comprehensible and manageable structure

>> machine

3. INTEGRATION PHASE

A holistic structure or system

>> organism

HOW MIGHT WE...
enshrine design thinking in the organization?

For organizations that have not yet gathered experience in design thinking, it can help to examine its excellence (its fitness!) in detail before beginning. If we develop design thinking exclusively at the initiative of one single area, it won't have any lasting effect. From our experience, it is promising to establish the basis for effective design thinking through a company-wide network of users and supporters. This way, design thinking can be disseminated transversally. The buy-in of the decision makers stays as an imperative, though. Management must invest in the development of the capabilities of the entire organization to create an integrated organization.

What do we need in order to implement an integrated approach in a company?

The ideal is that all employees see themselves as entrepreneurs and act accordingly. Because an integrated and customer-centric company is characterized by the fact that, alongside the company management, the organizational structure and the implementation processes are oriented toward the customers/ecosystem. All employees act on their own responsibility, and the work is meaningful to all involved.

We recommend paying attention to the following elements:

Company management

The company management should place customer centricity as a crucial, strategic theme in the organization and communicate it to all employees. Coupled with a clear vision, it thus empowers all employees to direct themselves fully toward the customer and the ecosystem. For the employees to be able to work independently and in a customer-centric way, management creates a basis of trust. From this basic attitude, a mindset can evolve that gears the strategy toward the attainment of the common goal: to be of service to the customer/ecosystem.

The organizational structure and organizational culture

The organization needs a correspondingly open structure and culture, which are characterized by collaboration. An atmosphere is created in which commitment and a focus on the customer/ecosystem can be lived and experienced. Such an organization facilitates networking and simultaneously a high level of autonomy for all involved. A culture emerges in which collaboration is radical and fast.

Holistic implementation of the customer experience

Customer centricity raises the awareness within the organization for a holistic implementation of the customer experience. To ensure the competitive edge and as a response to changing customer needs, it is pivotal that the whole organization act flexibly—that customer knowledge be quickly and iteratively integrated with improved customer experience chains, and that these experiences be shared with potential partners in the ecosystem.

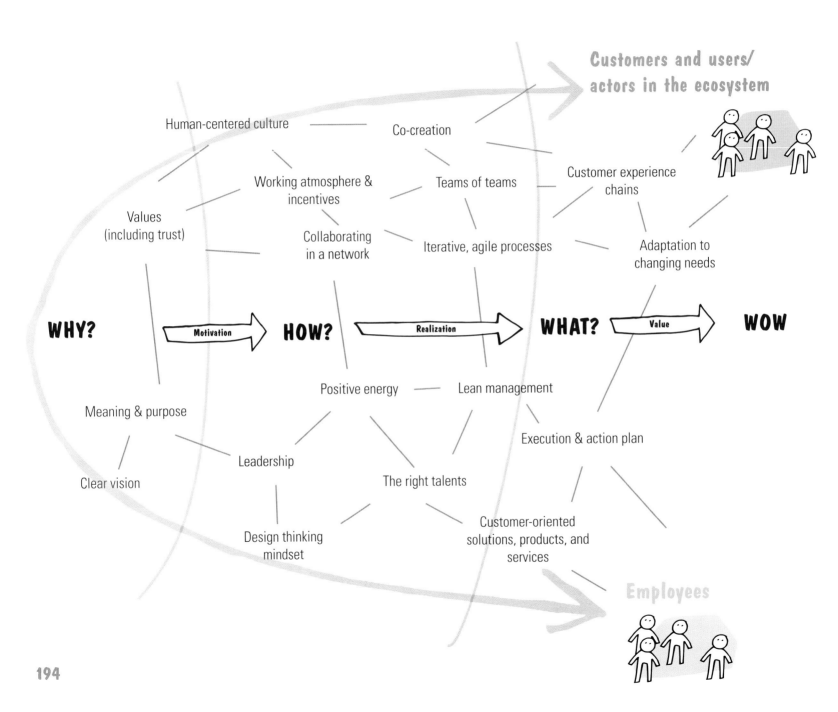

Customers and users/actors in the ecosystem

Human-centered culture

Co-creation

Working atmosphere & incentives

Teams of teams

Customer experience chains

Values (including trust)

Collaborating in a network

Iterative, agile processes

Adaptation to changing needs

WHY?   Motivation →   HOW?   Realization →   WHAT?   Value →   WOW

Positive energy

Lean management

Meaning & purpose

Execution & action plan

Leadership

The right talents

Clear vision

Customer-oriented solutions, products, and services

Design thinking mindset

Employees

For change management to result in an integrated and customer-centric organization, we can start with asking how advanced customer centricity is in the company. If customer centricity is still poorly developed, appropriate measures for improvement can be taken. For our example of the racing bike, this means: We build up our muscles purposefully where we need the strength and endurance for our Alpine crossing!

The maturity of customer centricity in an organization can be determined, for instance, by means of self-assessment. Because customer centricity is a feature of the organization as a whole, we should include all employees to determine its maturity. The employees usually know best where the organization can be improved and what should be improved.

The prevailing methods for finding out the degree of development of an organization are general assessments, for instance as part of the EFQM self-assessment or as part of a traditional employee survey. From our experience, approaches such as the Customer Centricity Score™ (CCScore™) are more suitable. In this approach, the focus on the customer is consciously chosen as the starting point of a measurement method. The CCScore™ measures the degree of dissemination of customer centricity in the company. The evaluation is then done on different levels of aggregation of an organization and allows for a differentiated view of how strong customer centricity is and, hence, how "fit" the organization is. On this basis, it can be established where we should begin with the development of the mindset.

Customer centricity

-30

Low score
"Customers are tolerated"

81

High score
"Customers are in the focus"

Customer centricity is measured mainly with the goal of inferring from the result any specific measures to boost it. Simply to know that we are not comprehensively focused on the customer does not yet change the mindset!

The measurement of customer centricity is similar to a survey on customer satisfaction. If we involve only a few people, the result is patchy. To get to meaningful information, we must question a representative cross-section of the workforce. Much more important than the measurement itself is the dedication that develops from it. Employees must be actively involved in the development of measures. This is the only way they develop their mindset further.

The measurement results only constitute the starting point of a multi-tier process that leads to the targeted improvement of customer centricity. In a closed circuit comprising measurement and inventory, reflection and the development of measures, as well as the subsequent implementation in the organization, causes and effects of improvement measures can be tracked and controlled.
More elements of digital transformation are discussed in Chapter 3.6.

Step 1: Measure the strength

The strength of customer centricity is measured by an online assessment. This allows for a very detailed and differentiated view of the individual drivers behind the CCScore™. It will become quite clear what the individual factors contribute to the overall score of the company and where there is potential for improvement.

Step 2: Infer specific options for action

In a method-based reflection, the causes and drivers of the CCScore™ results are analyzed, translated into relevant strategies for improving customer centricity, and written down. This so-called U procedure is basically a change process, which develops far more than just customer centricity; it enables the company to strengthen the effectiveness of its organization fundamentally and thus lays the cornerstone for effective design thinking.

Step 3: Define an action plan and implement it

To implement the strategies that have emerged, an action plan is drawn up for selected measures. The implementation of the measures is initiated; progress is regularly monitored; and goal attainment is checked in the following CCScore™ measurements. The development of customer centricity can be thus monitored and controlled.

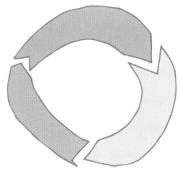

1. Measurement

2. Options for action

3. Implementation

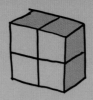

KEY LEARNINGS
Transform organizations

- Create an organizational structure without silos—this is the only way to disseminate design thinking transversally in the company.
- Establish a mindset that is focused on designing experiences (e.g., positive experiences across the entire customer experience chain) or "Wow!" effects when dealing with the product.
- Put the customer and his needs at the center of any activity. He is the reason the company exists.
- Consider customer centricity and design thinking as active, complementary aspects of change management. Live the "change by design" thought.
- Measure customer centricity (e.g., by way of an index) and improve it step by step.
- The transformation into a new mindset encompasses all levels: management, structure, and implementation.
- Raise the awareness of company management for a new mindset. Create commitment to and confidence in the new way of working throughout the organization.

2.7 Why strategic foresight becomes a key capability

Peter, Priya, Lilly, Jonny, Linda, and Marc have one thing in common. They are all on Facebook. This makes them six amid more than one billion Facebook users worldwide. Facebook has grown into the largest social network in less than a decade. Facebook's mission is to provide their users with the possibility of sharing information and thus creating an open and networked world. Mark Zuckerberg's way of thinking in terms of implementing the strategy is straightforward: "We go mission-first, then focus on the pieces we need and go deep on them, and be committed to them." The great success of Facebook is based on long-term thinking, in particular on a strategy planning with a perspective that goes beyond five years. As soon as the strategy process is completed, Zuckerberg begins to break down the strategy into small and implementable sub-projects (missions) for his teams.

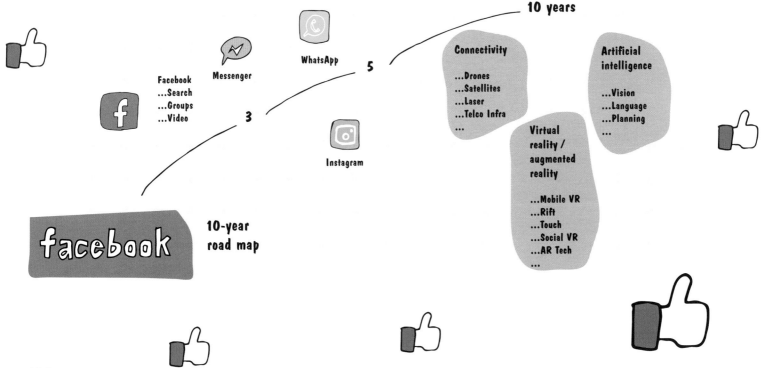

10 years

5

3

Facebook
...Search
...Groups
...Video

Messenger

WhatsApp

Instagram

Connectivity

...Drones
...Satellites
...Laser
...Telco Infra
...

Artificial intelligence

...Vision
...Language
...Planning
...

Virtual reality / augmented reality

...Mobile VR
...Rift
...Touch
...Social VR
...AR Tech
...

10-year road map

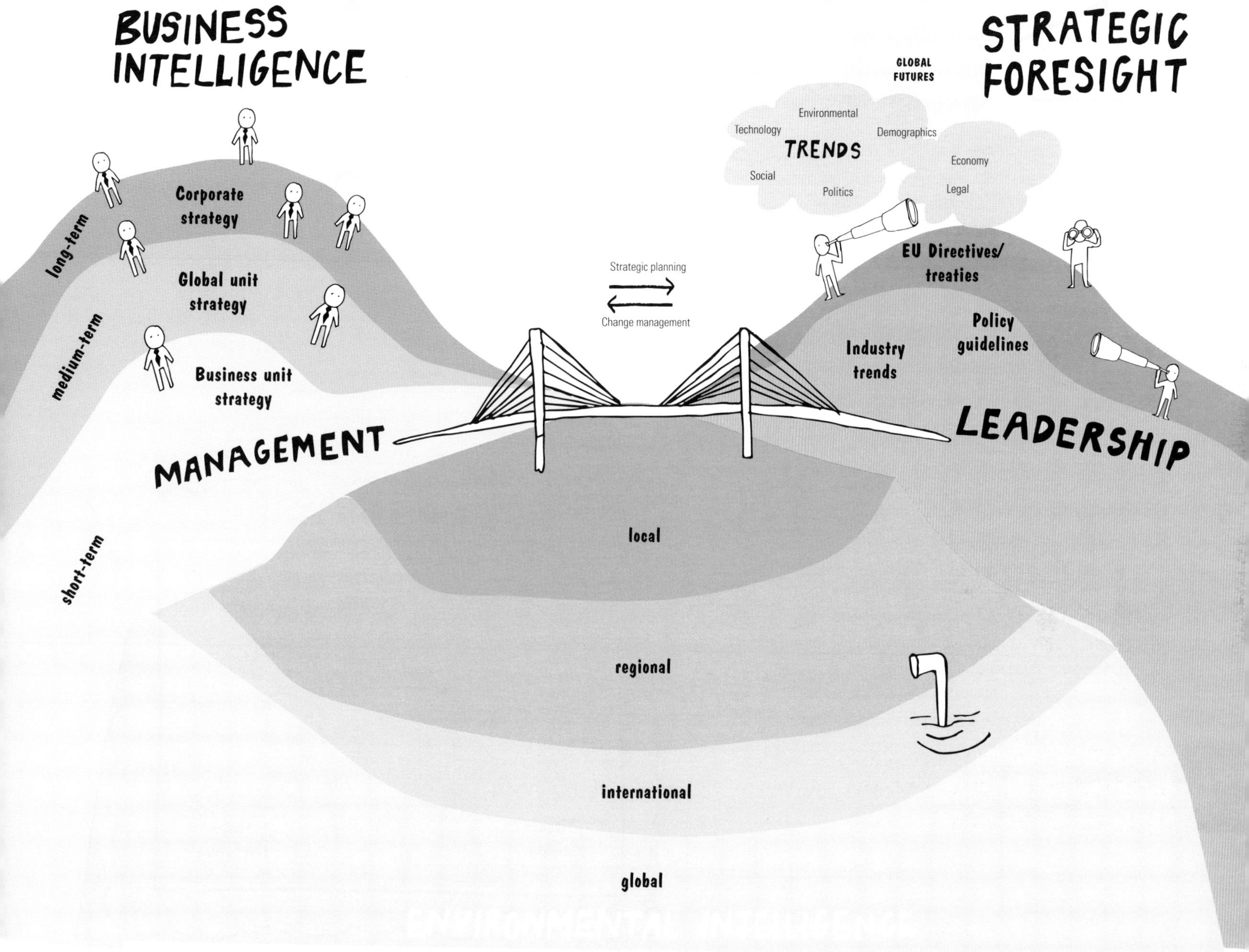

BUSINESS INTELLIGENCE

STRATEGIC FORESIGHT

GLOBAL FUTURES

TRENDS

Technology
Environmental
Demographics
Social
Politics
Economy
Legal

long-term

medium-term

short-term

Corporate strategy

Global unit strategy

Business unit strategy

MANAGEMENT

Strategic planning

Change management

EU Directives/ treaties

Industry trends

Policy guidelines

LEADERSHIP

local

regional

international

global

199

How do business leaders such as Zuckerberg develop products and services for the customers of tomorrow?

Their method is referred to as strategic foresight, an approach that focuses on shaping the desired future. Strategic foresight consists of a mindset and a methodology in tune with it. The mindset is characterized by a belief in the future, which we can shape ourselves by hunting for new market opportunities. The methodology includes various tools and techniques that lend a hand in steering the teams systematically and purposefully in the right direction.

The Strategic Foresight Framework was developed at the Stanford Center for Design Research and has evolved into four schools represented by Drucker, Schwartz, Jouvenel, and Arnold. Of course, the Framework is embedded in design thinking.

For the development and change of the culture, the mindset is of overriding importance in order to realize the opportunities the future offers. A mindset consists of established attitudes, values, and opinions, which are currently brought to day-to-day work. In the last few years, various mindsets have been discussed that attempt to make the motivation and behavior of people graspable. With a "foresight mindset," we are convinced we have the future in our own hands. This future is translated step by step into reality with targeted activities. In a company, a foresight mindset can form the basis for launching a new business area or developing an innovative product.

Peter Schwartz
"Scenario planning as a basis for the development of stories about the future."

Peter Drucker
"Strategic planning as a basis for decisions that are made today about the future."

STRATEGIC FORESIGHT

Bertrand de Jouvenel
"Visions as a basis for human action in the context of society, the economy, and politics."

John Arnold
"Multidisciplinary approach for processing complex problems with the aid of future scenarios that go beyond the power of the imagination."

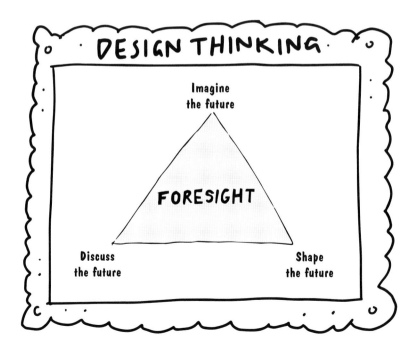

What does WYSIWYG stand for?

The idea of **"What you see is what you get"** became popular in the late 1960s—mainly known by its abbreviation WYSIWYG (pronounced: WIZ-ee-wig) . It means we accept as given what we see before us. It can be compared to the preview of software in which the HTML code is already visible in the visualized interface. A few years ago, the wisdom of **WYFIWYG** took hold at Stanford: **"What you foresee is what you get."** Thus we found out that expectations have an impact on what will happen in the future. What we foresee and intuit also influences the result!

Why is planning important?

For good planning, it is vital to know the effect of a possible future. This means we develop the ability to change our own opinion and that of the team.
Positive moods influence the team. By creating a positive attitude toward future market opportunities, we are better prepared for realizing them:

We can do better!
We are more effective!
We have a different approach to the solution of a problem!

Such a mindset can be learned and applied in each organization, each team, and during any growth phase of a company. At this point, it should be noted that this approach is very different from that of futurologists. Futurologists usually claim that they are able to map the future through scenarios and trend analyses. They base their forecasts on past data or current trends, which they extrapolate into the future. The model for strategic foresight discussed here is based on different considerations. It combines a long-term perspective with the well-known tools of strategic planning and design thinking. This combination enables teams to tackle short-term fields of action that are aligned to medium-term and long-term market opportunities.

WHAT
YOU
(FORE) SEE
IS
WHAT
YOU
GET

Peter really likes the advanced mindset and sees in strategic foresight an ideal complement to his current activities.

But he also knows that, in the minds of many business leaders, the belief prevails that a sort of secret recipe must exist that enables companies to launch radical innovations on the market. There is a plethora of rumors of various kinds:

- Success is due to the genius of the company founder and serial entrepreneur (Apple & Tesla)
- Unlimited resources are the decisive factor (Google & Facebook)
- It's just pure luck (Twitter & Snapchat)

It's me ✗ aha! ✗ ♥ = Oⲙ

Countless books have been written on the gloomy future of companies that don't succeed in deciphering the secret recipe. Often companies have no clear idea as to where the journey should lead. Their excuse is that other companies don't have a clear strategy and vision either and that it is still too early to formulate a clear digitization strategy anyway.

Ultimately, a clear vision is the only ingredient that determines success. What spoils success **are prevailing mindsets and management paradigms based on fear.** Fear is easy to stir up, such as with negative financial forecasts about missed targets, cost savings, and the announcements of staff layoffs.

The good news is: There are other ways!

Over the last 50 years, a culture was created at Stanford and in Silicon Valley that allows for a team to act in a forward-looking manner and thus develop new products and even industry standards. New models from design and engineering research, and the lecture halls and labs in the Valley, became effective tools for global innovation leaders. Many of these tools were documented in the "Playbook for Strategic Foresight and Innovation," which is available free of charge to all innovation fans at www.innovation.io. The foresight mindset is disseminated virally in organizations when effective tools are applied accordingly and the long-term mindset is backed by all. Companies such as Deutsche Bank, Volvo Construction Equipment, Samsung Electronics, and many other global businesses have established a new mindset in their organizations with the aid of these tools. The Foresight Framework was developed to protect teams from a "disruption dread": the growing fear that others might overtake you. The distinguishing feature of the Foresight Framework is a positive vision of the future, the products, the services, and the vision of a company.

The Foresight Framework has such a simple structure that anybody can go through the five stages from beginning to end.

The first three phases are dedicated to the question of how we best deal with a new or hitherto unknown problem statement. We begin with the famous white sheet of paper and try to capture the future. The problem statement can be directed inward or outward. In extreme cases, it can even mean that we redefine the future of an entire enterprise.

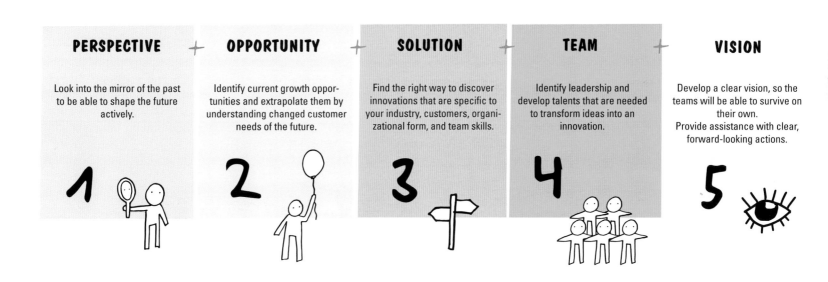

PERSPECTIVE + **OPPORTUNITY** + **SOLUTION** + **TEAM** + **VISION**

PERSPECTIVE
Look into the mirror of the past to be able to shape the future actively.

OPPORTUNITY
Identify current growth opportunities and extrapolate them by understanding changed customer needs of the future.

SOLUTION
Find the right way to discover innovations that are specific to your industry, customers, organizational form, and team skills.

TEAM
Identify leadership and develop talents that are needed to transform ideas into an innovation.

VISION
Develop a clear vision, so the teams will be able to survive on their own.
Provide assistance with clear, forward-looking actions.

In the initial phase (**"Perspective"**), the focus is on understanding the past. This reflection helps us to comprehend what has happened up to the present day. Once a team has understood the past, it is easier to understand possible options for the future. The question of why these options are rated high especially can yield the decisive insight when they are implemented later. Often project teams immediately address the solution. For this reason, it is a vital component of the Foresight Framework to go through this phase and recognize the added value gained in the reflection on the past.

The next phase (**"Opportunity"**) deals with understanding the needs of potential customers. When we recognize unfulfilled customer needs, we have already come closer to customer groups, who are most ready for changes, and we automatically address the question of how these customers might benefit from an innovation.

The third phase (**"Solution"**) concentrates on the building of prototypes as a potential solution to the problem statement. The value of the solution option that was developed can be better assessed when compared to other solution options and workarounds.

In the fourth phase (**"Team"**), the issue is to develop routines for the talents that will help them find new ideas and develop them further.

The last phase is focused on the **Vision**, which is indispensable for the viability of an idea. Only with a clear vision will the various stakeholders support the idea, invest their lifeblood in it, and ultimately contribute to scaling the idea.

The latter two phases help to enshrine strategic foresight throughout an organization. This process takes time because organizations do not learn overnight. This is why every phase has used three tools that are very useful when processing the respective questions (see p. 201). Studies conducted at Stanford have proven that people find such tools helpful for arranging and structuring their thoughts, especially when it comes to knowledge-intensive and complex issues. Strategic foresight is based on abstract ideas with a high level of uncertainty in the early stages of development. The tools help in the analysis of what is already known and what is not. The future can also be made palpable and graspable this way.

New ways of working and tools for a successful transition are crucial in the age of digitization, which entails great uncertainty for businesses. The ability to anticipate with strategic foresight seems to be of fundamental importance in the 21st century for managers and agile teams alike.

IT'S THE FUTURE!!!

EXPERT TIP
Integration of strategic foresight with design thinking

How can we develop a digital vision?

Good design thinking adapts to the situation, wins people over with a strong mindset, and helps us with the digital transformation of the company. Strategic foresight expands our view of the future and generates the great visions we need if we want to participate in the next market opportunities. Strategic foresight embeds design thinking in an infinite product and service continuum, which comes to life iteratively in our future-proof concepts. Thus it helps to develop a long-term perspective and robust ideas. The goal is to direct attention to potential opportunities and risks through integrative approaches and a networked way of thinking, and to infer appropriate conclusions from it. Through early detection, strategic foresight helps to deal with the high speed of change on the outside and the frequently prevailing internal inertia of the organization. Furthermore, it promotes the willingness to change.

How does strategic foresight support design thinking?

When we superimpose selected tools and methods onto the phases of research, conceptual design, and implementation, we get a "double diamond" with a vision in the center.

The methods of strategic early detection (strategic foresight) can be used quite well to sharpen the picture of the future, the vision, and future customer needs. It helps with the explanation, design, and selection (filtering) of important topics in the design thinking project. The set of methods in strategic foresight helps us with the definition of the digital vision.

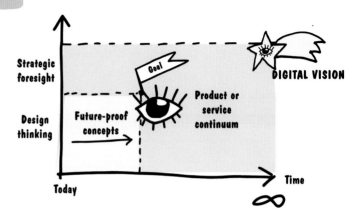

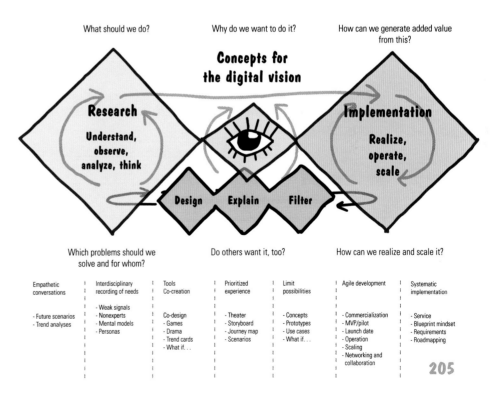

		What should we do?		Why do we want to do it?			How can we generate added value from this?	

	Which problems should we solve and for whom?			Do others want it, too?			How can we realize and scale it?	
Empathetic conversations	Interdisciplinary recording of needs	Tools Co-creation	Prioritized experience	Limit possibilities	Agile development	Systematic implementation
- Future scenarios - Trend analyses	- Weak signals - Nonexperts - Mental models - Personas	Co-design - Games - Drama - Trend cards - What if...	- Theater - Storyboard - Journey map - Scenarios	- Concepts - Prototypes - Use cases - What if...	- Commercialization - MVP/pilot - Launch date - Operation - Scaling - Networking and collaboration	- Service - Blueprint mindset - Requirements - Roadmapping

Progression curves help us to put events, life cycles, and other developments in the proper context (analogous to the S-curve model).

Janus cones support us in depicting multiple, overlapping, and intersecting events in a framework.

Change paths give us an indication of the most important milestones we need to reach in order to perform a certain action.

White spot analysis provides us with insight into hidden market opportunities and allows a wider view of the competitor landscape.

Buddy checks help us obtain a good match in terms of the right partners and team members.

Crowd clovers support teams with the mapping of innovation networks in order to make an idea ultimately come alive.

Alongside the concept of the future user (see Chapter 1.1), *generational arcs* can be used that illustrate demographic transformation and enable us to see things from the point of view of different generations.

A *real theater* allows us to immerse in the world of tomorrow and experience the needs of users in realistic environments (cf. storytelling, Chapter 2.4).

A *vision statement* is the clear and concise summary of an idea. It helps, for instance, to describe the prototype briefly and concisely (see POV, Chapter 1.6).

The *pathfinder* shows us the best path an idea should take through the organization or an innovation network (cf. stakeholder analysis, Chapter 3.4).

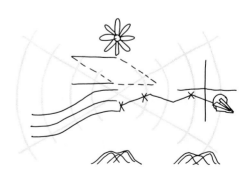

Build your own toolbox!

PROGRESSION CURVE

Examples of scenarios revolving around digitization, technologies, and business models. Through the combination of several trends in the progression curve, new, innovative ideas can be constructed (see example in circle).

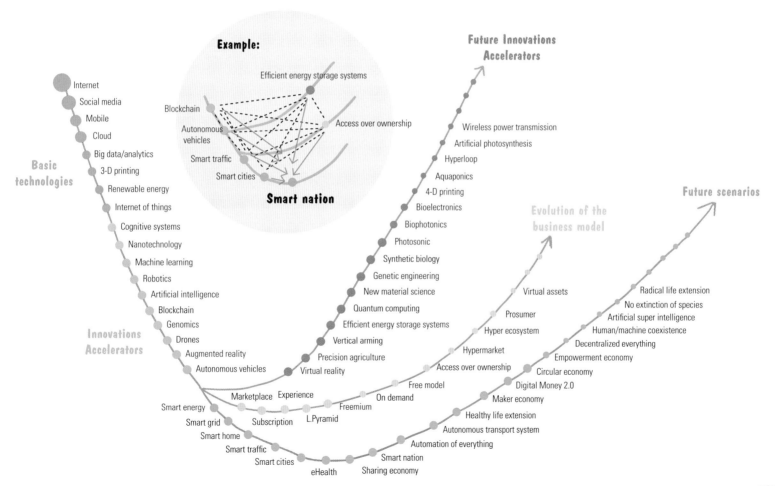

207

Application example: Combining foresight and design thinking to shape a vision of future mobility

By combining the mindsets, we can develop tangible concepts of the future that are custom-tailored to the user. The feeling of ownership—which is currently still a strong feature (e.g., possession of cars) but will very likely change due to new mobility concepts—can be depicted in a scenario about future mobility (through personas, needfinding, etc.). However, this transition becomes a reality only when the infrastructure of cities is geared to the new concepts, offering the user the best possible experience. On the basis of movement data (big data/analytics), it is possible to determine the optimal locations for concepts in terms of Car2go and Bike2Go, for bus stops and point-to-point connections, for instance. Zones and driving lanes for private mobility service providers (e.g., Uber, Lyft, Didi Kuaidi, etc.) and public mobility services (bus, train, tram, etc.) are also inferred from it. Large parking lots are turned into green areas; the old parking spaces along the streets become the new waiting lines for autonomous vehicles. By actively making possibilities for parking scarce in the cities and optimizing a range of offers of limitless mobility, the quality of life in the cities will increase in the medium term. Thus smart mobility will be a relevant pillar of smart city initiatives, which are connected to the respective mobility concepts through sensors, smart video surveillance, and video analytics with the goal of controlling such complex systems. Whereas models such as "access over ownership" provided only a weak trend signal only a short while ago, they have now developed into a macro trend that was picked up by many companies and affects numerous areas. An emerging micro trend is "decentralization of everything"—as can be seen in the first blockchain initiatives or free-floating parking spaces. From Singapore to Berlin, everybody is talking about the mega trend of smart cities.

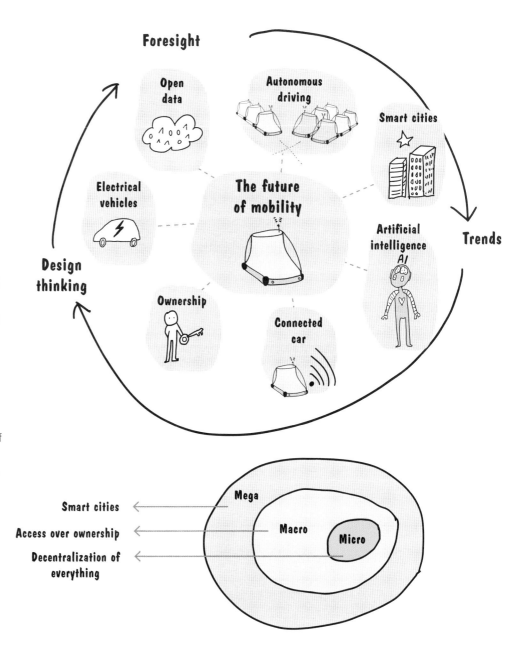

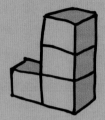

KEY LEARNINGS
Apply strategic foresight

- Apply strategic foresight as the ability to plan and shape the desired future.
- Understand the past in order to be able to shape the future.
- Recognize and extrapolate the changes in customer needs.
- Take a positive view of the future and use tools to make this future become a reality.
- Work with interdisciplinary teams to achieve a transversal dissemination of the mindset in the company.
- Develop a clear vision so that everybody on the team pulls in the same direction.
- Define clear-cut steps so the vision can be implemented in a targeted manner.
- To do so, use the tools described in the "Playbook for Strategic Foresight and Innovation."
- Develop a combined mindset from design thinking and strategic foresight for the digital vision.

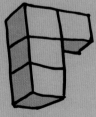

3. DESIGN THE FUTURE

We begin our chapter on designing the future with systems thinking, although the approach and the mindset are at least as old as the design thinking paradigm. We are firmly convinced, though, that the basic conditions and interaction of systems must be taken into account more and more when we develop our future products, services, and business ecosystems. The use of a converging mindset of systems thinking and design thinking will be pivotal in many areas.

The last time Peter dealt with systems engineering was during his time as a student at Munich Technical University. He can recall quite well a discussion during a lecture in the context of the *Challenger* explosion on January 28, 1986. It was determined at the time that the system had not been adapted to safety needs, and this was why the terrible disaster occurred. Peter often thinks about the disaster. How complex are things when a self-driving car is on the road? How many systems must interact with one another?

Engineered systems have a reason for their existence: They implement a desired or required function. For example, we want to build a self-driving car for a stress-free drive from point A to point B. As an alternative, we can integrate the autonomous vehicle in a system of means of transportation and won't ever again have to search for a parking space, because the vehicle will be permanently be on the road as part of a larger system. For this, the responses from certain sensors and information in the vehicle are important for communicating the necessary parameters to the system on how it must adapt to its environment. A rain and cold sensor, for instance, in combination with a camera or radar can provide information on road conditions and thus be an indicator for the speed to be chosen. To achieve this, all components must interact. With regard to self-contained technical systems, complexity is manageable. But as soon as nature as such and our social systems come into play, forecasts are far more difficult. Traffic will increase when we no longer park our autonomous vehicles but have them circulate in the cities. It's our own motives in a system that are difficult to explore and comprehend.

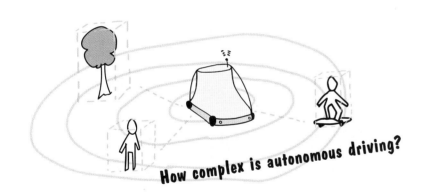

How complex is autonomous driving?

Many things can be understood as systems: products, services, business models, processes, and even our family or the organization in which we work. We use the term "system" to describe the interaction of several components (system elements) in a larger unit and its environment. All these elements fulfill a specific function or a purpose. In what follows, we use the terms "systems thinking" and "systems engineering" synonymously to a large extent.

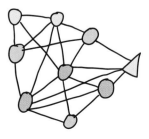

The tools and methods from systems thinking that go beyond drawing up and creating systems help us to model, simulate, and later produce complex systems in a future human–machine and machine–machine relationship—especially if we want to solve wicked problems with design thinking and are faced with the challenge of capturing the environment with its ever-growing complexity. Examples of complex systems are: coral reefs, nuclear power plants, or our introductory example of autonomous driving.

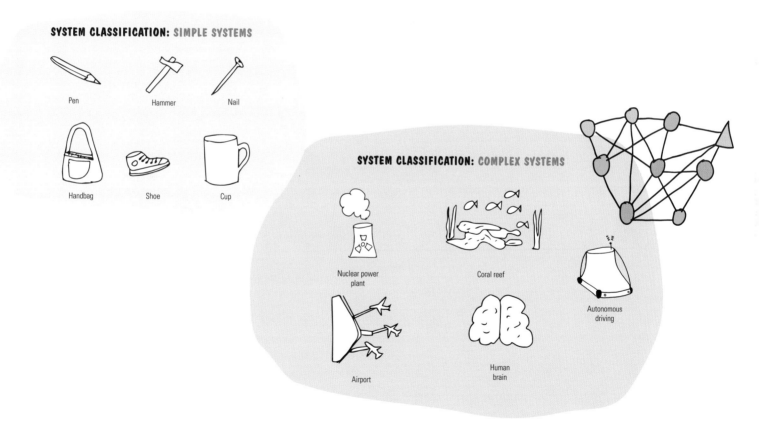

SYSTEM CLASSIFICATION: SIMPLE SYSTEMS

Pen

Hammer

Nail

Handbag

Shoe

Cup

SYSTEM CLASSIFICATION: COMPLEX SYSTEMS

Nuclear power plant

Coral reef

Autonomous driving

Airport

Human brain

How is the modeling done (mapping of reality)?

Delimitation of systems is a central task of modeling. Especially because effectiveness and efficiency today are more important than ever for the development of new systems. It is obvious that the error probability of complex systems is greater than that of its individual elements. With the use of modules and sub-elements and the introduction of redundancies, we attempte to reduce the probability of failure of the system as a whole.

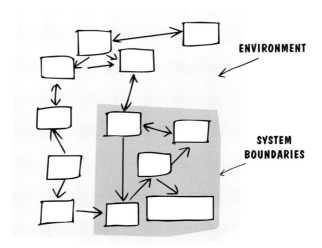

This is based on the assumption that we can influence and change the elements within the system boundaries. The elements within the system boundaries are the strengths and weaknesses known to us. The elements outside these boundaries are the opportunities and risks that affect our system.

What are the elements a systems thinking process consists of?

Put simply, systems thinking is another problem-solving method that uses a variety of elements to optimize the system.

Response and feedback are vital elements of systems thinking. Unlike linear models, which consist of cause/effect chains (A causes B causes C causes D, etc.), in system thinking the world is seen as a connecting unit with various relationships (A causes B causes C causes A, etc.).

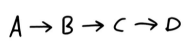 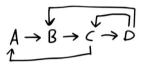

The advantage of a model with feedback is that it does not just map what happens at what time, but yields information on how something happens and why it happens. In this way, we learn how a system behaves. Over time, feedback loops increase the response; it can go in both directions: positive and negative. For this reason, it is important to stabilize the feedback loops. Using the feedback only for the optimization of the gap between the target state and the actual situation is a good way of stabilizing.

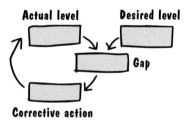

When we deal with the implementation of systems, we must ask ourselves five core questions:

- Which gaps affect our systems, and to what degree?
- Do we know the gaps and are we able to describe them?
- How do we monitor the gaps?
- What possibilities do we have to close the gaps?
- How great is the effort to close the gaps, and how much time do we have to do so?

How does systems thinking work?

In systems thinking, a specific initial problem from the real world (1) marks the beginning. With complex problems, the real world is usually multidimensional, dynamic, and nonlinear. In a first step, we try to understand the system and map the reality (2). This mapping, or system representation, helps us to understand the situation (3). The situation analysis is about comprehending the situation step by step—from rough to detail. We can use various methods here, such as mathematical models, simulations, or experiments, and prototypes. We summarize the findings of the situation analysis in a SWOT analysis, for example, on the basis of which we formulate the goals (4) to be fulfilled by the solution. This way, we obtain the decision-making criteria for the assessment of the solution.

The situation analysis is important for finding out where there are still gaps with regard to the target state. At this point, improvements are usually still necessary, or we simply still lack information to close the gap.

Only once the problem and the situation are really known do we begin with the search for a solution (5). It is now important to identify solutions that actually do fit into the solution space.
In this phase, we endeavor to find several solutions (i.e., to think in variants). By way of synthesis and analysis, we generate different solutions, which we evaluate in the next step (6).

We apply decision-making criteria to the evaluation. Tools and methods such as evaluation matrix, logical argumentation, simulations, experiments, and so forth, have proven their effectiveness.

Based on the evaluation, a recommendation is given and a decision is made (7). If the solution meets our requirements and solves the problem, that's good; otherwise, we iterate the process until we have solved the problem completely.

In systems thinking, a strong focus is on the continuous communication with the stakeholders. This means that their consent can be obtained at an early stage during critical phases of the development. The output of our representation can be documented as the operational concept (ref. ISO/IEC/IEEE 29148).

PROBLEM SPACE

SOLUTION SPACE

7 Decision
Recommendation

6 Evaluation
e.g., model, simulation, experiment . . .

Criteria

"Reality"
dynamic
multidimensional
nonlinear

Inputs

1 Initial problem

2 Mapping of reality

4 Goal formulation (Decision-making criteria)

Synthesis & analysis

5 Search for a solution
Solution that fits into the solution space. Thinking in variants

3 Situation analysis
From rough to detail
Mathematical models and simulation
- Experiments and prototypes
- Abstract thinking
- Data analytics
- Summary of findings (e.g., SWOT)

Solution variants

What mindset does a systems thinker live?

Systems thinking is an interdisciplinary approach whose primary goal is solving complex problems or implementing technical systems that depend heavily on each another. As mentioned, the system is divided into subsystems. The individual elements are specified and processed. In so doing, the entire problem (e.g., across the entire life cycle) and the technical, economic, and social framework conditions of all customers or stakeholders should be taken into consideration. Systems thinking offers a team-oriented structured approach for doing so.

A good systems thinker, therefore, masters different ways of thinking and concentrates correspondingly on the requirements on hand. He switches the perspective, such as from individual parts to the whole, or from structures to processes.

We always have our eyes on the big picture.

We think positively of a way to improve the system and don't complain when it doesn't work.

We check the results and improve the result with every iteration.

We reflect on our way of thinking because it affects what will happen.

We take the time to penetrate even complex interconnections.

MINDSET OF A SYSTEMS THINKER

We search for the "key" to the system.

We consider facts from various perspectives.

We accept that change takes place gradually and that interconnections also trigger changes.

We identify the effects that are triggered by an action.

Where and how do the design thinking and the systems thinking mindsets converge?

The mindsets of design thinking and systems thinking have some similarities; differences are of a complementary nature, so the convergence of the two approaches is quite exciting.
What both paradigms have in common is the goal of better understanding the problem and the situation. To achieve this goal, we work on interdisciplinary teams, using different methods and tools. It is important that the team always knows where it is in the process and that it acts in a goal-oriented way. Visualization and modeling are factors of success in both approaches.

The similarities are:
- Coverage of the same or similar thematic areas.
- The purpose and goal is the solution of (complex) problems and the simultaneous definition and expansion of the solution space.
- It's important to clarify the critical variables and functionalities at the onset of the project to reduce risks.

From the terms used so far, we quickly realize that the focus of systems thinking is on the system, while the focus of design thinking is on the human being, the user. Both paradigms use a clearly defined but differently aligned problem-solving cycle as well as an iterative approach. Iteration in systems thinking aims at gradual refinement; in design thinking, many iterations enable us to understand the situation better and to approximate a potential solution.
By combining systems thinking and design thinking, the combined application of systemic, analytical, and intuitive models of thinking is also supported—and thus the finding of holistic solutions.

Systems thinking	Complementary mindset	Design thinking
Focus on the system	Different focus	Focus on users and needs of people
Systematic analytical problem-solving cycle	Clearly defined but different (problem-solving) process	Intuitive, circular problem-solving cycle
White box view with a focus on solution space	Design and architecture of systems	Black box view with focus on problem statement
Gradual refinement of the system	Iterative procedure	Carry out a great many iterations quickly

Systems thinking	Similar mindset	Design thinking
Create clarity through the consideration of the system and changes over time	Create clarity	Create a common understanding and clarity
Establish clear structures and anticipate life cycle considerations	Process understanding important (be mindful of process)	Process understanding is important
Mapping and modeling of the system	Visualization	Visualization and prototyping are important
Use of methods from systems thinking	Use of various tools and methods	Use of methods from design thinking
Collaboration and the exchange of information with stakeholders is pivotal	Interdisciplinary collaboration on the team	Initiate radical collaboration
System understanding helps reduce uncertainty	Positive dealing with uncertainty	Perform experiments in order to learn
Conduct project management in a target-oriented way	Focus on action	Implementation-oriented and solution-oriented action

HOW MIGHT WE...
use systems thinking in design thinking?

We don't want to get into philosophical speculations here as to whether design thinking is superimposed on systems thinking or whether the processes should be ranked in a hierarchy. From our experience, it is best when design thinking and systems thinking complement each other as the situation requires it.

If we take a typical development process as a basis, we can assume that design thinking is a strong tool in an early phase (conception and feasibility). This is especially true when the issue is simple functionalities or the interaction with a potential user. For the interaction of components, the simulation of complex processes, or the engineering of requirements, systems thinking is predestined for many developments.

Thus design thinking can help not only during various phases in the development process, it also contributes a number of factors and mental attitudes, which are usually not part of systems thinking:

- Arriving at new solution approaches that are brilliant in their simplicity
- Focusing on systems-in-systems with an alignment to individuals or entire groups (360 degrees) in terms of empathy
- The iterative approach in relation to the building of simple prototypes during problem solving
- Doing it and not planning a long time

The combination of the two mindsets results in new opportunities and better problem solutions!

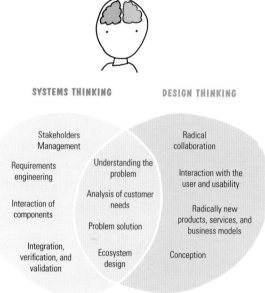

From the point of view of a design thinker, the way of thinking about systems and system boundaries in different situations can be helpful; for example, not just for a real, in-depth, and clear understanding of the problem space and solution space, but also for the identification of so-called blind spots and relationships between actors or for the generation of new ideas.

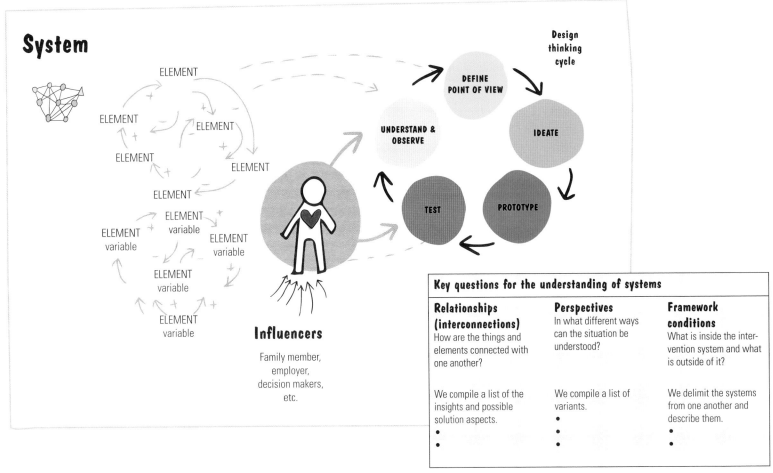

System

ELEMENT

ELEMENT

ELEMENT

ELEMENT

ELEMENT

ELEMENT

ELEMENT

ELEMENT
variable

ELEMENT
variable

ELEMENT
variable

ELEMENT
variable

ELEMENT
variable

Influencers

Family member,
employer,
decision makers,
etc.

Design thinking cycle

UNDERSTAND & OBSERVE

DEFINE POINT OF VIEW

IDEATE

PROTOTYPE

TEST

Key questions for the understanding of systems

Relationships (interconnections)	**Perspectives**	**Framework conditions**
How are the things and elements connected with one another?	In what different ways can the situation be understood?	What is inside the intervention system and what is outside of it?
We compile a list of the insights and possible solution aspects.	We compile a list of variants.	We delimit the systems from one another and describe them.
•	•	•
•	•	•

HOW MIGHT WE...
apply systems thinking and design thinking in tandem?

As mentioned, the switch from systems thinking to design thinking and vice versa can help to alter one's focus and perspective. With this switch, we change our focus from a product-centric to a people-centric approach.

It makes us design thinkers more aware that we ourselves are a part of a system in its environment. With our actions, we affect the entire system; we can intelligently interact with it; but we also realize that other stakeholders/observers might have a different view of the system as a whole. The system of a family is a good example. We know the actors of our family. Living together consists of complex interactions, and we have the possibility of changing the system through our actions. In addition, people who do not belong to our family have a different perception of our clan than we have inside the family.

System: The Jones family

Communication
Cohesion
Relationship

Why should we take their point of view?

Systems thinking helps us to identify effective actions with the system. Our ability to learn is strengthened, and we build on the basis of human thinking when designing our systems. In addition, the system can have higher cognitive skills.

The basic questions posed to the system environment are:

1.   What does the system produce? Is the result desirable?
2.   How does the interaction of the system with us as human beings work? Does the interaction correspond to our needs?
3.   What happens within the system? How do machines and sensors interact with one another? What do we want to achieve?

We recommend using systems thinking in tandem with design thinking in the case of complex (wicked) problems. How strong the combination of the mindsets should be depends on the project requirements or personal preferences and should be adapted accordingly. To design thinking experts such as Peter, we recommend switching the thinking mode and trying out the system-oriented problem-solving cycle, especially in phases of stagnation or when the overall picture is unclear.

In the case of a strong personal stamp with the systems thinking mindset, it is as important to confirm the findings at least once by means of the design thinking problem-solving cycle and to expand the creative framework. In most cases, this adds new insights, which can only find their way into the problem solution in an intuitive problem-solving cycle. On closer examination, the two approaches are not so different after all. Both follow the double diamond model and switch between divergent and convergent ways of thinking.

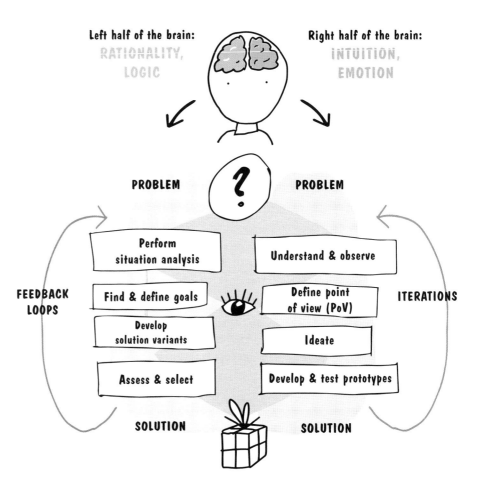

KEY LEARNINGS
Understand complexity with systems thinking

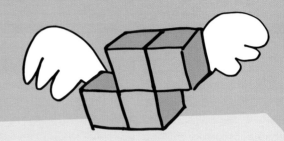

- Define the system, system boundaries, and factors that influence the system.
- Map out the relationships—within and outside the system boundaries.
- Make sure that all stakeholders in the system are listed.
- Look at the problem holistically; assume that it is complex per se.
- Even a complex, multidimensional, nonlinear, dynamic problem can be mapped and modeled as a system in a simplified way.
- Start with the simple stuff and go from rough to detail.
- Begin with the search for a solution once you have understood the problem (or aspects thereof).
- Always think in variants when searching for a solution.
- A graphical system representation helps to understand the problem and communicate the solution.
- Consider your first system picture to be a prototype that will be continuously tested and improved.
- Use a wide range of concepts, methods, and tools.
- Take different views and perspectives. Only one view is surely the wrong one.
- Combine systems thinking and design thinking into a common mindset.

In a dynamic environment, there is little time to plan activities over a long period. The lean start-up mindset is therefore the best for the continuation of design thinking activities. Like design thinking, it focuses on short iteration cycles and the consideration of customer feedback. At the end, the product life cycle and the development of the business model should be designed in such a way that few costs are incurred.

We have had good experience with the lean canvas by A. Maurya (see part A on next page), which can be expanded very well by other design thinking methods, such as customer profile (see part B) and the experiment reports (see part C) by A. Osterwalder.

The canvas is like the blueprint of an architect. It outlines the key factors, which ultimately cover the most important areas of a company: problem, solution, customers, value propositions, and financial viability.

For the start-ups of Lilly and Marc, it is crucial to define a unique selling point. Lilly is faced with the challenge of having to define what exactly differentiates her start-up from the countless other consultancy firms on the market. Marc must find features that convince patients to manage their "patient record" on his solution. Furthermore, the value proposition is designed to provide a solid basis in the long run for monetizing the collected data.

Thus the unique selling proposition (USP) constitutes the core of the canvas that either solves a specific problem of the customer (e.g., sovereignty over patient data) or meets a specific need (e.g., design thinking consultancy adapted to Asian business practices).

On the right side of the canvas, the sales channels and revenue are entered alongside the customer segments (divided into target segment and early adapters).

The left side of the canvas concentrates on rational reasons. The focus is on the problem statement, your own solution, and the existing alternatives, as well as on the cost structure. Supplements to the customer profile help us to understand the needs of the customer better. The reports of the experiments document our approach and show the progress per iteration.

TOO FAT

TOO THIN

LEAN START-UP

Lean canvas **+** **Customer profile**

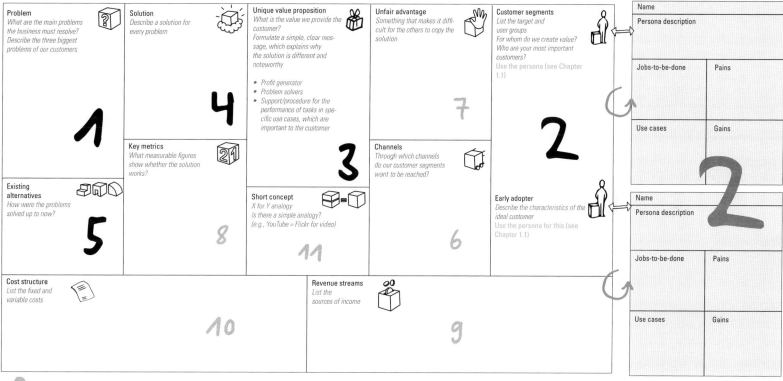

Problem
What are the main problems the business must resolve? Describe the three biggest problems of our customers

1

Solution
Describe a solution for every problem

4

Key metrics
What measurable figures show whether the solution works?

Unique value proposition
What is the value we provide the customer? Formulate a simple, clear message, which explains why the solution is different and noteworthy

- Profit generator
- Problem solvers
- Support/procedure for the performance of tasks in specific use cases, which are important to the customer

3

Unfair advantage
Something that makes it difficult for the others to copy the solution

7

Channels
Through which channels do our customer segments want to be reached?

Customer segments
List the target and user groups For whom do we create value? Who are your most important customers? Use the persona (see Chapter 1.1)

2

Early adopter
Describe the characteristics of the ideal customer Use the persona for this (see Chapter 1.1)

6

Existing alternatives
How were the problems solved up to now?

5

Short concept
X for Y analogy Is there a simple analogy? (e.g., YouTube = Flickr for video)

8

11

Cost structure
List the fixed and variable costs

10

Revenue streams
List the sources of income

9

Name
Persona description

Jobs-to-be-done	Pains
Use cases	Gains

2

Name
Persona description

Jobs-to-be-done	Pains
Use cases	Gains

+ **Experiments reports**

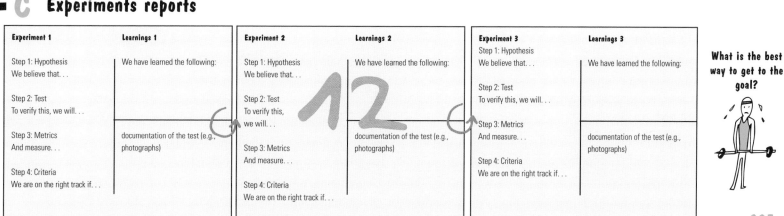

Experiment 1

Step 1: Hypothesis
We believe that. . .

Step 2: Test
To verify this, we will. . .

Step 3: Metrics
And measure. . .

Step 4: Criteria
We are on the right track if. . .

Learnings 1

We have learned the following:

documentation of the test (e.g., photographs)

Experiment 2

Step 1: Hypothesis
We believe that. . .

Step 2: Test
To verify this, we will. . .

Step 3: Metrics
And measure. . .

Step 4: Criteria
We are on the right track if. . .

Learnings 2

We have learned the following:

documentation of the test (e.g., photographs)

12

Experiment 3

Step 1: Hypothesis
We believe that. . .

Step 2: Test
To verify this, we will. . .

Step 3: Metrics
And measure. . .

Step 4: Criteria
We are on the right track if. . .

Learnings 3

We have learned the following:

documentation of the test (e.g., photographs)

What is the best way to get to the goal?

HOW MIGHT WE...
develop the lean canvas step by step?

Typical start at 1. or 4.

4. Solution

MVP/MVE

Question:
What is the idea/solution?

Methods:
- Creativity techniques
- Solution interview
- Analogies/benchmarking
- . . .

3. Unique value proposition

Question:
What is our value proposition/what is our unique selling proposition (USP)?

Methods:
- Value proposition design
- NABC
- . . .

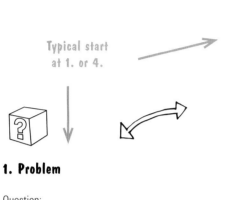

1. Problem

Question:
What is the (main) problem?

Methods:
- Problem interview
- Stakeholder map
- 6 WH
- . . .

2. Customer segments

Question:
What do the customer profiles look like?

Methods:
- Customer segmentation
- Customer profile
- Persona
- . . .

11. Short concept

Question:
Is there a simple analogy?

Methods:
- Business model analogies
- Creativity techniques

5. Existing alternatives

Question:
How has the problem been solved up to now?

Methods:
- Interview & observation
- Competition analysis
- Best practices
- . . .

226

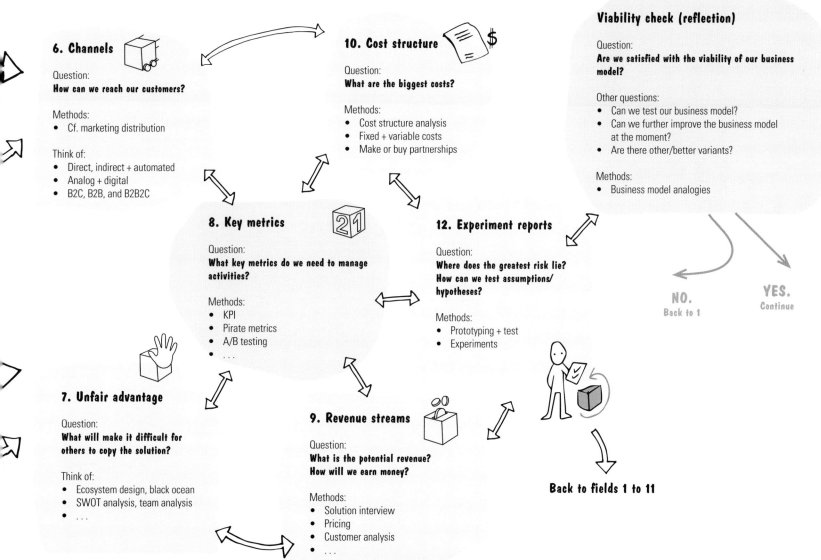

6. Channels

Question:
How can we reach our customers?

Methods:
- Cf. marketing distribution

Think of:
- Direct, indirect + automated
- Analog + digital
- B2C, B2B, and B2B2C

10. Cost structure

Question:
What are the biggest costs?

Methods:
- Cost structure analysis
- Fixed + variable costs
- Make or buy partnerships

Viability check (reflection)

Question:
Are we satisfied with the viability of our business model?

Other questions:
- Can we test our business model?
- Can we further improve the business model at the moment?
- Are there other/better variants?

Methods:
- Business model analogies

8. Key metrics

Question:
What key metrics do we need to manage activities?

Methods:
- KPI
- Pirate metrics
- A/B testing
- ...

12. Experiment reports

Question:
Where does the greatest risk lie?
How can we test assumptions/hypotheses?

Methods:
- Prototyping + test
- Experiments

NO.
Back to 1

YES.
Continue

7. Unfair advantage

Question:
What will make it difficult for others to copy the solution?

Think of:
- Ecosystem design, black ocean
- SWOT analysis, team analysis
- ...

9. Revenue streams

Question:
What is the potential revenue?
How will we earn money?

Methods:
- Solution interview
- Pricing
- Customer analysis
- ...

Back to fields 1 to 11

As described, the classic entry point into the lean canvas in an innovation project is via the problem (or oftentimes the solution); then the value proposition is inferred via the customers (by means of the customer profile) **(version A).**

With digital business models in particular, the entry can be made via other areas as well:

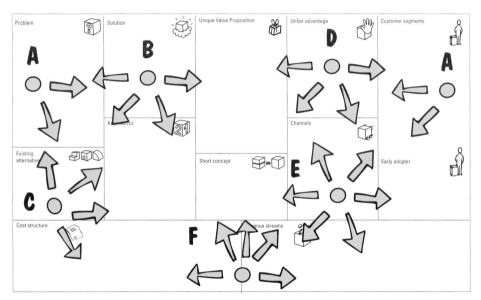

B) via the solution

We look for physical objects that have not yet been digitized and create the same customer application digitally instead of with a physical object.

Example: iPod/CD -> streaming services or 3D printers for food

C) via missing existing alternatives

At present, almost everything is being digitized. It is safe to assume that many things can be digitized. We look for things for which there is no digital alternative yet.

Example: Fiat money -> cryptocurrencies

D) via the creation of an unfair advantage

An unfair advantage as a protection against one's own business model being copied can also be an entry point, such as by building up an entire business ecosystem.

Example: WeChat (see Chapter 3.3)

E) via digital instead of physical channels

Companies such as Amazon continue to expand their channel access with ever new services. Digital access, directly or through a partner, can be the entry point.

Example: Amazon services

F) via existing business models

Existing business models that have not yet been implemented in one's own environment or language area. Existing solutions are copied and adapted to the new digital context.

Example: Online retailers for eyeglasses

Often, the business model becomes the differentiating factor per se. New business models may aim at securing income in the long term, heightening customer loyalty, or reducing production costs. One trend that has prevailed is servitization. The term means that customer needs are fulfilled without the customer having to own the physical product.

One of the best-known examples is Rolls-Royce's power-by-the-hour model in the area of aircraft engines. The pay-per-use model in the software industry as well as business models in the textile industry are also popular. Socks or baby clothing can now be ordered with a monthly subscription. Vigga has developed a model that enables mothers to lease organic baby clothing.

To stay with the textile industry: For another business model, the issue of recycling is the decisive differentiating element. The basic idea is to generate value from a waste product. Flippa K takes back the worn clothing of their customers, for instance, and sells it in their own secondhand shops.

The efficient use of resources can be an important part of the business model. Companies like YR are typical of this idea. It makes it possible for customers to create their own personal T-shirts, which are then produced on demand. This way, there won't be any overproduction of designs, which are later hard to sell on the market. Consumers also have a greater emotional bond to the product and the brand.

NEWBORN

Circular value creation	Servitization	Frugality
Create value from waste – close the cycle	Create value for customers without having them to own the product (physical good)	Promote the effective use of resources
Recycling and resource efficiency are in the focus	All-inclusive carefree packages, pay-per-use models, etc.	Solutions that actively lower consumption and production
Flippa K operates second-hand shops	Rolls-Royce: power by the hour	YR: On-demand production of T-shirts and shoes
Closed loop	Renting & leasing	Co-creation

♔ EXPERT TIP
Develop value proposition

The need/approach/benefits/competition (NABC) analysis is another simple tool for drawing up a value proposition. If you are able to answer the key questions from the four areas adequately, the value proposition is usually clear.

Unlike the customer profile, the NABC analysis takes the competition in due consideration, so it allows for an additional focus on uniqueness.

The needs and benefits are derived from the customer profile. The approach corresponds to the solution, and the competition is in line with the existing alternatives in the lean canvas. The value proposition is derived from these elements.

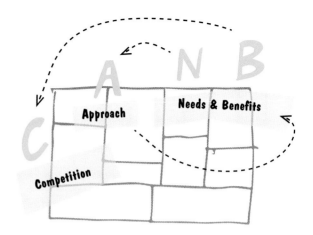

| **N**EED | **A**PPROACH (Solution) | **B**ENEFITS | **C**OMPETITION (Existing alternatives) |

NEED	**A**PPROACH (Solution)	**B**ENEFITS	**C**OMPETITION (Existing alternatives)
• What customers do we address? (internal/external) • What is the main customer need? • What does the customer have problems with? • Where are opportunities for improvement? • Where does our opportunity lie? • What are the main problems?	• What is the approach to a solution or the performance promise? • What is the product, service, or process proposal? • How will the product or service be developed and launched on the market? • How will we earn money with it? (business model) • What technological drivers affect our business model?	• What are the benefits for the customer? • What are qualitative and quantitative benefits for the customer? • How can I communicate them in the form of storytelling?	• What alternatives exist today and in the future? • What is the risk? • How were the problems solved up to now?

What makes up a good value proposition?

From our experience, there are 10 key factors of success, which we should absolutely check during the validation of the value proposition:

1. Embedding in a great business model.
2. Focus on what is important for most customers.
3. Focus on topics for which the customer is willing to spend a lot of money.
4. Focus on unresolved problems.
5. Aims at only a few tasks, pains, and gains but solves them extremely well.
6. Alongside the functional task fulfillment, it takes into account the emotional and social components of the tasks.
7. Alignment to the customer's measurement of success.
8. Differentiation from the competition.
9. Better than the competition in at least one dimension.
10. Difficult to copy.

How do we communicate a value proposition?

The value proposition we communicate should be formulated as concisely as possible in one brief sentence. This helps in communication and makes the business idea clearer by way of an analogy, such as "Sailcom" (car sharing for boats) or "WatchAdvisor" (TripAdvisor for watches). This analogy is described in the lean canvas as the "short concept."
Lilly still has trouble formulating her concept in one sentence. She tries out "Book and Fly for Design Thinking" or "Last Minute Problem Solving." But she is not really happy with her ideas.

EXPERT TIP:
With service design thinking to value excellence

When we have already made our company fit for the future through operational and service excellence, service design thinking provides the necessary framework to initiate differentiation. As noted, business models with a strong focus on servitization may come into being, or customers may be segmented by categories that follow a different rationale than age, income, and family status. Service design also includes experience design, UX, and UI. Ultimately, the basic considerations are based on design and systems thinking.

The insurance industry provides us with an example of the application of service design. The Swiss health insurer Sanitas integrates the Swissmom portal in their service portfolio. Thus a customer interaction chain is created with expectant mothers, from the initial wish for a child, to pregnancy, to giving birth, up to dealing with toddlers. The actual product of insurance protection for newborns is embedded in the service components. In addition, mothers are given access to a community and a further range of offers for childcare.

The quality and operational excellence in the service used to be in the foreground. Then, more and more importance was put on experience. Today and in the future, value will take center stage. Value excellence is achieved through a strong customer centricity and close, proactive collaboration with customers. So-called customer experience chains constitute the basic structure; the touch points with the customer from the initial interaction to the warranty case are mapped in the chain.

From our experience, we can recommend two tools:

- **The customer journey (customer experience chain)**
- **Service blueprint (an extension of the customer experience chain as a blueprint)**

The customer experience chain, or customer journey, represents the process that the customer experiences in contact and dealing with our company. The issue is to design the journey of the customer through our offers. All contact points (touch points) should be considered.

The service blueprint is an extension of the customer experience chain and also constitutes the provision of the service. On page 234, we will address the topic of using a service blueprint.

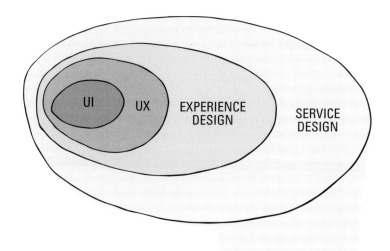

UI UX EXPERIENCE DESIGN SERVICE DESIGN

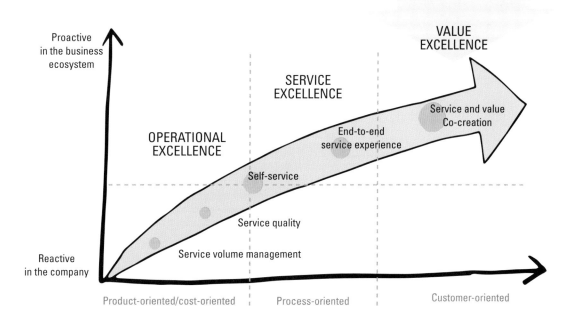

Proactive in the business ecosystem

VALUE EXCELLENCE

SERVICE EXCELLENCE

OPERATIONAL EXCELLENCE

Service and value Co-creation

End-to-end service experience

Self-service

Service quality

Service volume management

Reactive in the company

Product-oriented/cost-oriented

Process-oriented

Customer-oriented

233

HOW MIGHT WE...
use a service blueprint?

A service blueprint can be used to describe, for instance, a service prototype in a uniform, systematic, and structured way.

A service blueprint is a method for visualizing and structuring service processes. Various levels are differentiated:

• Activities the customer goes through

• Visible activities of the provider with respect to the customer

• Invisible activities of the provider that are directly connected but are not visible to the customer

• Supporting activities and systems of the provider or his partners

The customer interaction line separates customer activities from provider activities.

The method is easy to understand, easy to apply, and puts the customer in the center. The blueprint can be iteratively improved and adapted. The iterations also help identify the weak points in the process.

From our experience, teams can easily create a service blueprint using Post-its. We can go through the following steps:

1. Actions and activities of the customer on the timeline (large and small circles for the respective activity)
- What are the main actions?
- What happens in this step?

2. Customer interactions (green Post-its)
- What are the touch points with the customer?

3. Visible activities of the provider (blue Post-its)
- Who are the actors?
- Who is involved all together?
- What are the actions of the provider?

- What are the activities that are invisible to the customer?

5. Supporting activities and systems (orange Post-its)
- What supports the whole thing (software, platform, processes)?

6. Evaluation (red Post-its):
- What is critical? Where are errors possible?
- Where are risks and vulnerabilities?

Afterward, we can infer ideas for improvements, new processes, or entire service innovations. The service blueprint also helps us in interviews/tests with customers/users to capture the situation and obtain feedback.

Customer experience chain

visible

invisible

EXPERT TIP
Build the bridge from a concept to a scaling solution

Normally, our design thinking journey ends with the concept. We have already successfully demonstrated desirability, technical feasibility, and economic viability. But we are still a long way from scaling the solution. In the next phases, business design and product development are frequently done separately; from our experience, it is profitable to create the development paths for the customers, the business, and the product in close collaboration.

The iterative **service development**, for example, can be done in close contact with the customer. With customer experience chains, service blueprints, mock-ups, test Web sites and apps, and so forth, it is easier than ever to perfect services iteratively and in ever faster cycles with the customer. **Business design** can also be done in close contact with the customer. The business model is tested, adapted, and refined. All the other components, from marketing up to the value proposition, can be tested and developed with lead users or potential customers. Good business design is characterized by the fact that an obvious market opportunity is identified and transformed into a scaling business.

Furthermore, customers must be developed and won over during the development phase in order to increase the probability of a scaling business in the end. Steve Blank calls this approach **customer development.**

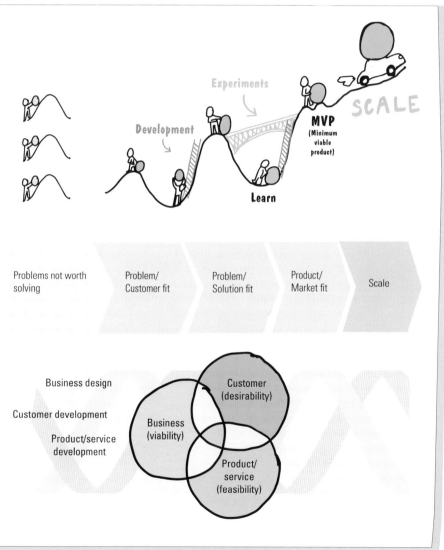

HOW MIGHT WE...
proceed in a structured way to achieve scaling?

With design thinking, we achieve a customer/problem fit—that is, our understanding of the customer and the problem has the necessary depth. Using lean start-up, we initially create a problem/solution fit; then we refine it into a product/market fit. Over the course of the process, we reduce the risk step by step through experiments; at the same time, we heighten the value of the project. From the simple prototype, we develop a minimum viable product (MVP), expand it with each iteration, and test it with the customer. To launch and execute successfully the design of business ecosystems becomes of paramount importance. An MVE helps to test the desired ecosystem scenario.

Because the various approaches build on one another, we charted the development steps with their potential approaches again by way of example. The steps apply especially to innovators such as Marc, people who are still at the beginning of their journey and have identified the problem to be solved.

It goes without saying that the different approaches, such as lean start-up, business design, and customer development, which all have a mindset similar to design thinking, can be combined with one another.

Described next are the individual steps of the integral problem to growth and scale framework.

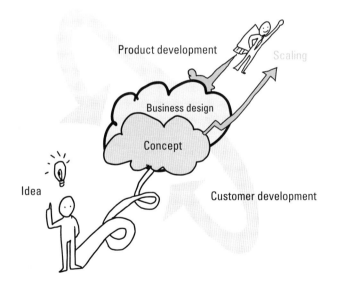

Product development

Scaling

Business design

Concept

Idea

Customer development

PROBLEM TO GROWTH & SCALE FRAMEWORK

1 Design thinking

- Determine your potential users, customers, and stakeholders
- Identify the real customer needs with design thinking
- Find solutions that are as elegant as they are simple
- Use systems thinking and data analytics

3 Co-creation

- Win more customers, users, and lead users and retain them
- Get the necessary help from the outside
- Work on teams across departmental and organizational boundaries
- Develop MVP/MVEs and build up trust in partners and customers

5 Business ecosystem design, agile product and customer development

- Shift your activities from problem solving and search for a solution to finding the right business model with business ecosystem design
- Develop the product and the business model further with agility (e.g., with methods like Scrum)
- Think in variants when developing business models
- Multidimensional consideration of the business models of all actors in the ecosystem becomes a success factor

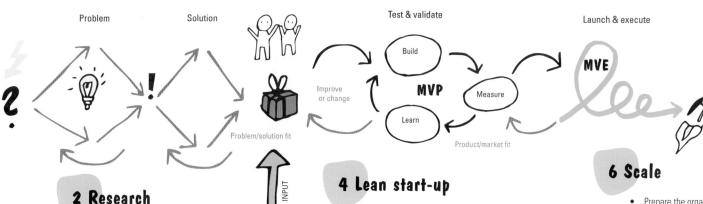

2 Research

- Understand the problem and the situation holistically
- Take advantage of market research instruments
- Validate and supplement your findings

4 Lean start-up

- Use the lean start-up approach to develop your offer further with little capital
- Structure the solution step by step
- Improve and validate your business model with fast iterations
- Clarify the biggest uncertainties with experiments

6 Scale

- Prepare the organization for growth and scaling
- Establish scalable processes, structures, and platforms
- Check the mindset and skills in your organization and don't just follow a blueprint
- Bring the entire organization one step forward and break new ground

As soon as we have reached an initial scaling, lean management helps us to keep the structures lean and make the most of our potential. It is therefore an important element to keep our innovations alive and a sort of fuel for the scaling.

From our experience, it is useful to expand the mindset of design thinking with its customer orientation by the following principles for existing products and services. Especially in fast-growing companies, the entire value creation chain must be mastered.

- We focus on our own strengths.
- We optimize our business processes constantly.
- We bank on continuous improvement.
- We live internal customer orientation.
- We rely on the responsibility of the teams for the fulfillment of their mission.
- We act in decentralized and customer-oriented organizational structures.
- We offer the best support in the leadership of our employees.
- We exercise open and direct communication.
- We conserve resources and avoid waste.

Lean management supports the implementation of the ideas in the business model at various levels. For example, fixed costs can be reduced by outsourcing or hierarchy levels can be reconsidered by the expansion of competencies.

Typical tools of lean management are value stream design, continuous improvement process (CIP), 5S, TPM, kanban, and Makigami.

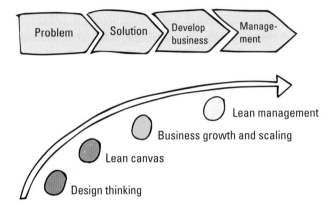

In addition, design thinking offers the methods needed to redesign processes and procedures by means of iterations. From our experience, it is of great value to make the process tangible, to shift process steps physically and scrutinize them. In this way, various activities along the entire value creation chain of the company were already improved. Ultimately, there are two levers: optimizing structures and processes, and scaling the excellence of teams. This requires at least T-shaped people, positive energy, and leadership.

KEY LEARNINGS
Design business models

- Start with design thinking to capture customer needs and get to the problem/solution fit.
- Use the lean canvas to summarize the findings from design thinking.
- Create different business model and lean canvas variants of your idea and select the most promising.
- Take your time: A good business model rarely emerges within a few hours; valuable insights are gained over the course of time.
- Determine a very good unique selling point and value proposition. Use a variety of tools and experiments for it: for example, the customer profile or the NABC analysis for inferring the value proposition.
- Reduce the risk systematically through experiments and adapt the lean canvas.
- Modify your business model at a later point in time when the market requires it or the customers desire it.
- Combine design thinking, research, co-creation, and lean start-up approaches even at established companies for innovation projects.
- Consider lean management approaches early in order to optimize efficiency and effectiveness in the case of a high level of scaling.

3.3 Why business ecosystem design becomes the ultimate lever

There is nothing revolutionary about thinking in business ecosystems. In the 1990s, James Moore described business ecosystems as economic communities supported by various organizations and individuals that are in interaction—he compares ecosystems with organisms of the business world. These organisms develop their skills and strengthen their market role in the system over time; in so doing, they tend to gear themselves to one or several companies. Today, the evolutionary approach through digital ecosystems is also referred to as "black ocean strategy." Well-known representatives of ecosystem designs are Apple and Android—both companies have successfully created ecosystems for apps.

Virtuous design cycle

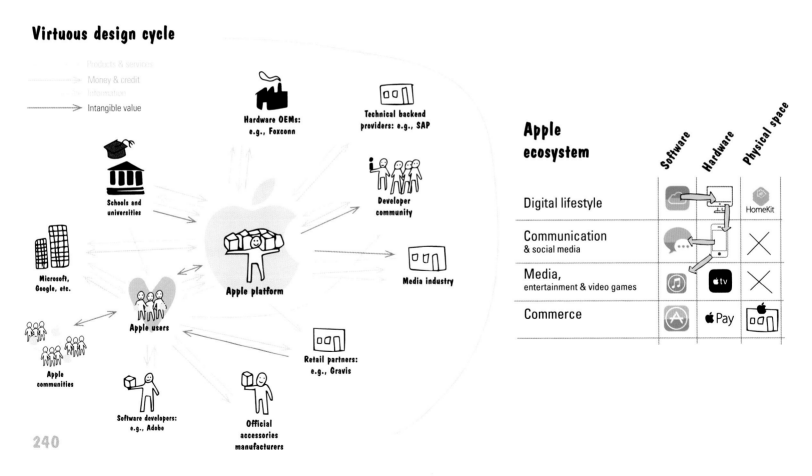

Another example is Amazon. Alongside its original core business, Amazon has succeeded in building several ecosystems, in which the company is operating successfully. They range from Amazon Vendor Express to Alexa/Echo all the way to Web services. Amazon is a good example of the effects of digital ecosystems. Such ecosystems integrate a range of digital offers under one brand, sell primary products, grow through targeted cross-subsidies of various services, and have open interfaces or ensure interoperability. In addition, lock-in effects are often produced, driven by a high level of user friendliness and security, coupled with claims to data sovereignty and security.

For the isolated consideration of a business model, it is sufficient to think in the "blue ocean" model (Kim & Mauborgne). A creative and cross-border redefinition of market services is key to this consideration, not least in order to differentiate yourself from the competitors. The goal of a black ocean strategy, however, is to make market entry impossible for the competitors.
Existing rules are changed, new framework conditions are created, and an "unfair advantage" is built up and correspondingly used. Systems thinking (see Chapter 3.1) and the design of business models (see Chapter 3.2) are basic skills for the design of such business ecosystems.

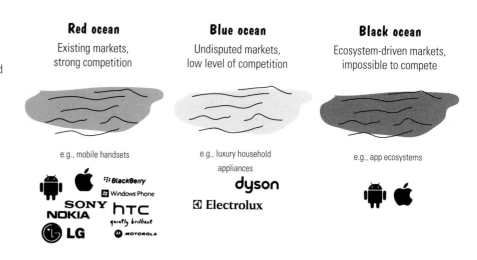

Red ocean
Existing markets, strong competition

e.g., mobile handsets

Blue ocean
Undisputed markets, low level of competition

e.g., luxury household appliances

Black ocean
Ecosystem-driven markets, impossible to compete

e.g., app ecosystems

Business ecosystem design as a paradigm for business models in distributed systems

Many projects in which blockchain is used as an enabler technology are a virtual Eldorado for business ecosystem design. The new, distributed networks cancel existing business models and make it possible to revolutionize processes, value streams, and transactions. Well-known models of business model design quickly reach their limits because they concentrate mostly on the primary business of a company and take only the direct customers and suppliers into consideration. The multidimensional view of the actors in the ecosystem with their value streams is often skipped. Thus thinking in business ecosystems in a business context becomes a factor of success.

What is the generic basic idea of a business ecosystem approach?

When building a business ecosystem, a pre-investment must be made. Costs are incurred, for example, to building a platform or other innovations for development of core skills. Establishing a platform shows that we have technical skills. However, this will not create long-term relationships with the other actors. From our experience, it is by all means recommended to invest in the business ecosystem as such, and expend some thought on how each individual actor in the system will benefit from our platform and which business models come to fruition for the respective actors. There will be actors who will lose their livelihoods in this environment on account of our ideas. The profit generated in the business ecosystem should cover these investments as long as other business models don't copy it. This brings us to the value of the overall offer in a simplified model. Ideally, the value of the overall offer rises steadily with the investments in the ecosystem.

What is important with digital business ecosystems is to think increasingly in decentralized structures (see diagram). They are not centralized customer-supplier networks in the traditional sense (maturity level 1), which are geared to one company or serve a linear customer experience chain. Centralized business networks (maturity level 2) are characterized by a central player, who tries to control the whole network. Such networks exist, for example, in the automotive industry. Digital business ecosystems often have no center, and many players act on an equal footing in the network (maturity level 3).

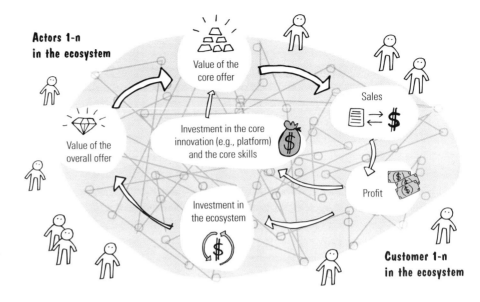

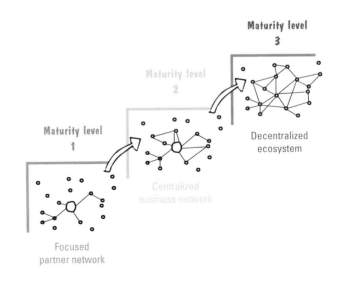

The concept of business ecosystem can be applied to all levels of maturity. In our considerations, we focus on the business ecosystem at the third maturity level, which has the following features:

- focuses on the user/customer
- loosely coupled and designed for co-creation
- networked and decentralized system elements
- coordinated and accepted value systems of the actors
- cross-industry offerings
- maximum benefit for the participants and actors
- enabled by new technologies (e.g., blockchain)

How do the ecosystem champions build systems?

The traditional path of an ecosystem structure usually consisted of iterations with a few customers to test the full value proposition. After the pilot phase came the rollout. The development costs were largely incurred before the rollout. An alternative procedure is visualized by way of the example of WeChat. The alternative path shows the evolution of a **minimum viable ecosystem (MVE)**. With this approach, the functionalities and the value proposition are heightened if there are enough players in the ecosystem. Today, WeChat is a digital ecosystem that has systematically developed on this path over the last 10 years. The ecosystem is currently expanding in terms of acceptance of cryptocurrencies and the integration of blockchain.

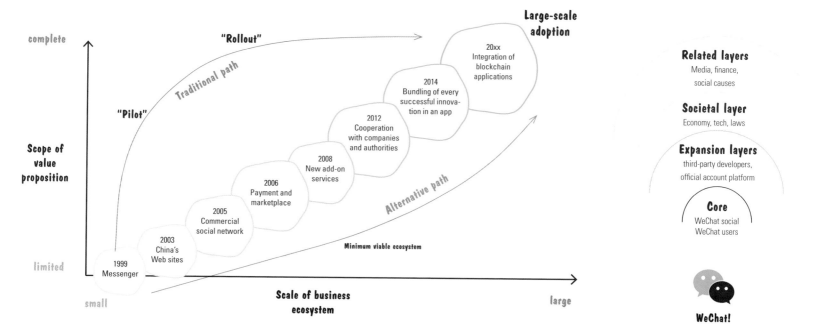

243

What does the business ecosystem map of Marc and his start-up team look like?

Marc and his team are facing an ecosystem design challenge for their digital business model in the health record environment, which they plan to realize based on a blockchain technology.
Marc has come up with a number of basic principles to shape his ideas for a business ecosystem. For one, he is convinced that blockchain as a technology will break through existing boundaries and industry-specific rules; secondly, he believes the dynamics in the industries will increase. Long-established industry giants in many segments will topple. This means there are market opportunities for a greenfield approach on all continents as long as a country possesses the necessary ICT infrastructure.

Marc and his team live by the principle that when you are fast and innovative, you will win in this game. When designing the business ecosystem, Marc always has his eyes on the big picture. He wants to design a system that makes it possible to establish the cycle of identification, treatment, service accounting, verification of services, medication, and reimbursement that delivers advantages to all actors and especially the patients in the system.

What are the needs of the users/patients?

Through observation and research, Marc and his team have found out a great deal about the health care system and the needs of patients. The team works with customer experience chains (see figure "Current patient journey"), which help to explain the interactions, needs, and disruptions in the system. Linda was able to contribute quite a bit of important information from the point of view of an expert because she knows the day-to-day hospital operation so well. Moreover, initiatives such as Google's Deep Mind have shown that such systems have potential. From this, Marc inferred the core value proposition, the basis for the further design of the ecosystem.

Current patient journey

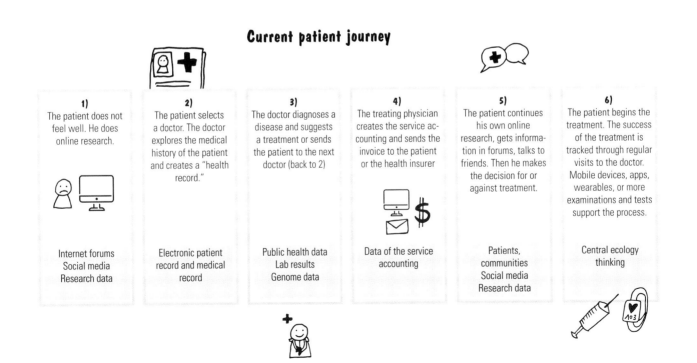

1)	2)	3)	4)	5)	6)
The patient does not feel well. He does online research.	The patient selects a doctor. The doctor explores the medical history of the patient and creates a "health record."	The doctor diagnoses a disease and suggests a treatment or sends the patient to the next doctor (back to 2)	The treating physician creates the service accounting and sends the invoice to the patient or the health insurer	The patient continues his own online research, gets information in forums, talks to friends. Then he makes the decision for or against treatment.	The patient begins the treatment. The success of the treatment is tracked through regular visits to the doctor. Mobile devices, apps, wearables, or more examinations and tests support the process.
Internet forums Social media Research data	Electronic patient record and medical record	Public health data Lab results Genome data	Data of the service accounting	Patients, communities Social media Research data	Central ecology thinking

Who are the users/customers and actors in the business ecosystem?

The start-up team has found various new needs from the point of view of the patients and the actors. However, the greater vision of the team is making the health care system more efficient, not least on account of the weaknesses of the exclusivity of the information in the invoices that often invites misuse. The start-up team wants to test its initial functions on the market with patients at an early stage and apply the lean start-up methodology.

Because the start-up team has the vision of improving the daily needs of the patients and actors in the system, thoughts are also focused on the service provider and the pharmaceutical industry alongside the patients. Taking these actors in due consideration, the business ecosystem is gradually redefined.

Where are the defined and other actors on the business ecosystem map, and which value streams are of relevance?

The start-up team puts the actors on the map and draws the different value streams. This way, the "virtuous design loop" for the MVE is created.

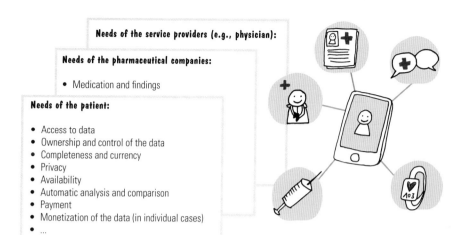

Needs of the service providers (e.g., physician):

Needs of the pharmaceutical companies:

- Medication and findings

Needs of the patient:

- Access to data
- Ownership and control of the data
- Completeness and currency
- Privacy
- Availability
- Automatic analysis and comparison
- Payment
- Monetization of the data (in individual cases)
- ...
- ...

Value streams

------> Products & services
------> Money & credit
------> Information
------> Intangible value
------> Digital assets
------> Cryptocurrency

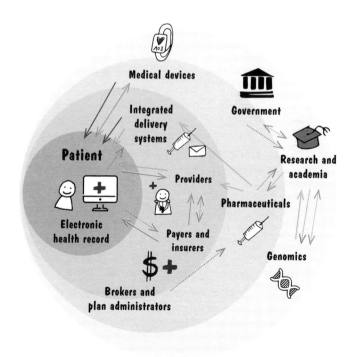

How are the different actors in the ecosystem integrated, and what advantages does the system offer the actors? Is the system accepted as an "organism"?

After several iterations on the business ecosystem map, the advantages for the three actors whom Marc and his team have in mind for their first MVE become visible. In the simplified model representation of the ecosystem, each actor has noticeable and strategic advantages, which come about through improved interaction with the patient. In addition, the team made a detailed analysis for each actor. The strengths, weaknesses, and opportunities and risks (SWOT analysis) as well as the resulting advantages and the motivation for each actor are therefore clear to the team.

How can the business ecosystem be redesigned? How do Marc and his team solve the efficiency and effectiveness problems in the health care system, addressing the needs of the patients?

A disruptive element of Marc's team is the implementation of the idea on a private blockchain for "health records." It includes all the patient's health information, and the patients decide with which actors they want to share health data. In addition, the data can be anonymized, so relevant information is filtered and knowledge about the efficiency and effectiveness of treatments is analyzed. Knowledge is created this way, and pharmaceutical studies are facilitated. The health data contains all the relevant information that is filed in decentralized systems. Access is through the access authorization of a private blockchain for which each patient has a "key." In a next step, the overall system is enhanced by means of artificial intelligence (AI) and machine learning to improve the effectiveness and efficiency of the health care system in the medium to long term.

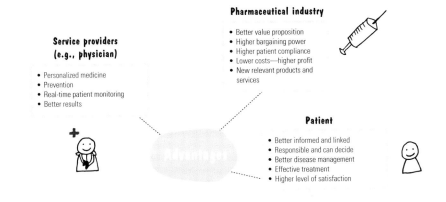

Service providers (e.g., physician)
- Personalized medicine
- Prevention
- Real-time patient monitoring
- Better results

Pharmaceutical industry
- Better value proposition
- Higher bargaining power
- Higher patient compliance
- Lower costs—higher profit
- New relevant products and services

Patient
- Better informed and linked
- Responsible and can decide
- Better disease management
- Effective treatment
- Higher level of satisfaction

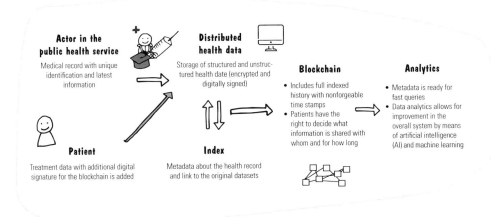

Actor in the public health service
Medical record with unique identification and latest information

Distributed health data
Storage of structured and unstructured health date (encrypted and digitally signed)

Blockchain
- Includes full indexed history with nonforgeable time stamps
- Patients have the right to decide what information is shared with whom and for how long

Analytics
- Metadata is ready for fast queries
- Data analytics allows for improvement in the overall system by means of artificial intelligence (AI) and machine learning

Patient
Treatment data with additional digital signature for the blockchain is added

Index
Metadata about the health record and link to the original datasets

How can Marc gradually build up the design skills in the business ecosystem and view the business models of all actors in all dimensions?

In preparation of a dominant role in the business ecosystem, Marc lives the design thinking mindset and proceeds iteratively. He used as the starting point the current structures in the business ecosystem, which he had observed. They had been established over the years on the basis of regulation and existing technology. By means of some design principles, such as the elimination of intermediaries or system providers, such ecosystems can be redesigned.

HOW MIGHT WE...
create a business ecosystem?

The central starting point in business ecosystem design is the customer/user with his needs, based on a defined problem statement. We use our well-known design thinking tools, such as customer experience chains, customer profiles, and personas. This is done before the design of the ecosystem. The design of the ecosystem usually takes place on two levels: customers/users and business, including the associated technologies and platforms. Our business ecosystem model has a total of 10 stages, which are broken down into a "virtuous design loop," a "validation loop," and a "realization loop."

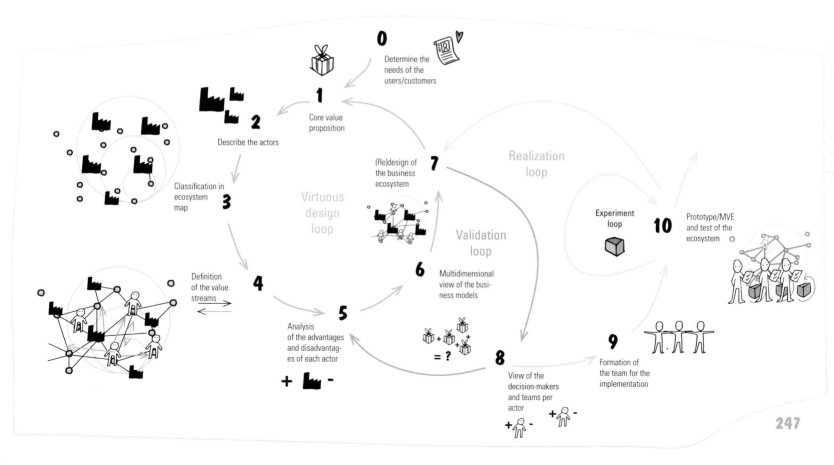

0 Determine the needs of the users/customers

1 Core value proposition

2 Describe the actors

3 Classification in ecosystem map

4 Definition of the value streams

5 Analysis of the advantages and disadvantages of each actor

6 Multidimensional view of the business models

7 (Re)design of the business ecosystem

8 View of the decision-makers and teams per actor

9 Formation of the team for the implementation

10 Prototype/MVE and test of the ecosystem

Virtuous design loop

Validation loop

Realization loop

Experiment loop

How do we start in the virtuous design loop?

1) Determine the core value proposition

The core value proposition for the user/customer, or for the system, is inferred from the customer needs.

2) Determine and describe the actors in the business ecosystem

Initially considering which actors could have relevance in the ecosystem is one good way to do this. There are a number of generic market roles in systems that we can define in advance. For the analysis, we can use well-known strategic and systemic methods of analysis (e.g., PESTEL analysis). Short descriptions of the companies, in which the function and role in the system, the primary motivation, and compatibility with our core value proposition are entered, help us summarize the findings. In addition, we note the intensity of the relationship and the current business model of each actor and other aspects.

3) Arrange the actors in the different areas of the ecosystem map

Enter the actors in an ecosystem map. For the business ecosystem map, we can work with a four-part division, for example; depending on the sector and use case, other structurings are possible. We place the core value proposition in the middle. The expanded, complementary offers and the enabling networks with their actors and their customers can be placed on the outer circles. The boundaries between the individual areas are blurred.

4) Define the value streams and connect the actors with the value streams

A core element in the business ecosystem design is the shaping of current and future value streams. For simple ecosystems in traditional businesses, we would be fine with physical product/service flows, money/credit flows, and information. For digital and digitized value streams, intangible values are highly relevant. Intangible values can be knowledge, software, data, design, music, media, addresses, virtual environments, cryptocurrencies, or access and transfer of ownership and possession. These value streams are increasingly decentralized and are exchanged directly between the actors. In addition, we should also bear in mind that there are negative value streams in the system, which emerge, for example, through a transfer of risk.

5) Create awareness of the advantages and disadvantages of each actor

After the actors are positioned in the ecosystem and clarity exists about the value streams, the effects can be analyzed for the individual actors. In this phase, we focus on the advantages and disadvantages that every actor gets from collaboration in the network. Without clear advantages, we will not be able to induce enthusiasm for the system in the actors.

6) Multidimensional view of the business models of all actors in the target business ecosystem

The analyses from the previous phases help us in the multidimensional view of the business models. We consider in particular the value proposition of each individual actor for his customers and, ultimately, what the actor contributes to the core value proposition for the customer/user. We make sure that the value propositions of individual actors match. In the end, all actors should perceive the distributions of opportunity and risk in the system as fair, and they should understood the value streams resulting directly or indirectly from the system. For many companies, the interaction with a digital business ecosystem is part of the digital transformation. In Chapter 3.6, we deal with this challenge again separately.

7) (Re)design of the business ecosystem

In this phase, the business ecosystem is iteratively improved. Actors are added in the iterations or are eliminated. For example, platform providers, hardware vendors, or value-added services can be added that change and improve the existing system. The impact on the individual actors and value streams should be determined for each variant or idea of the new or adapted ecosystem. From our experience, it is important to prove the robustness of the scenarios by means of iterations and experiments.

8) Look at the decision-makers and potential team members in the business ecosystem

We designed the system in phases 1 to 7. But only reality shows whether our ideas are really viable. In the validation loop, we consider with what specific actors we initially want to validate and develop our system. The so-called interspecific relationship between participating individuals and teams ensures the existence of a business ecosystem. This involves understanding the personal interests, needs, and motivations of those involved. Especially in a symbiosis (in the wider sense), in which all individuals benefit from the interaction, positive effects are generated, which lead to the growth of the system. Alongside the rational decision to be part of the ecosystem, personal motivation (e.g., from a decision maker) is at least as relevant.

What happens in the realization loop?

9) Form a motivated team for the design of the new business ecosystem

When designing business ecosystems, we have taken into account the needs of customers/users and of the actors. For a successful implementation, we also need the people who create the business ecosystem. The decision makers set the framework conditions, such as range of the MVE, budget, time frame, and so on. They are the enablers of the projects. The teams are the actual doers, who contribute positive energy, intrinsic motivation, interest, and skills.

10) Build the business ecosystem step by step with an MVE

Use the design thinking mindset and the approaches of lean start-up and agile development to build the ecosystem iteratively and improve it. Create prototypes and test them systematically. The redesign of ecosystems of maturity level 3 (i.e., those that effectuate a radical change in the market and revolutionize entire industries) constitutes a challenge to traditional businesses with regard to digital transformation. The corporate culture, the lived mindset, and the ability to think in business ecosystems are therefore critical to success, alongside the elements already described.

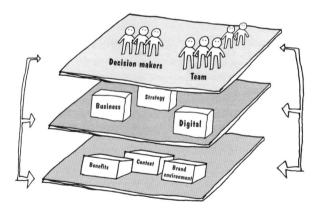

Because in many ways we work with canvas models (e.g., lean canvas or user profile canvas), we have had good experience in using the business ecosystem canvas (Lewrick & Link) for the iterative development of the system. The eight elements will help the design team across the entire design cycle (explore, design, build, test, redesign) in asking the right questions. In principle, the starting point for the design of a new ecosystem can occur anywhere. However, we recommend remaining consistently faithful to our typical starting point in the exploration phase, namely recording the customer and user needs.

In the ecosystem design canvas, all essential steps are consolidated. It is better we record the working version after each iteration (e.g., with a photo). This way, the considerations can be well documented and made traceable and comprehensible. In general, new systems (greenfield approach) or existing ecosystems can be improved with the business ecosystem canvas. When designing radically new eco-systems, certain actors in the business ecosystem can be eliminated already in the run-up. Another practical approach is first to draw up the business ecosystem predominant today and, in a second iteration, to optimize it (redesign). Especially if existing business ecosystems are to be radically restructured, the second approach makes sense, because in this line of thought, processes, procedures, information, and value streams can be redefined.

Business ecosystem design canvas

Determine the needs of the users/customers

- Who is the customer or user?
- Describe the customer/user profile (pains, gains, jobs-to-be-done, and use case)
- What problem is to be solved?

Core value proposition

What is the core value proposition for the user/customer?

Definition of the value streams

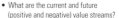

- What are the current and future (positive and negative) value streams?
- Which product/services streams, money/credit streams, data, and information flow?
- What are the digital and digitized value streams/assets?

DESIGN/REDESIGN

DESIGN:
- Which actors are pivotal for the provision of the core value proposition in the business ecosystem? (For the placement, go from the inside to the outside.)
- Also place the actors with advanced and complementary offerings, enabling functions and other actors who are directly or indirectly part of the system.

REDESIGN:
- Do various scenarios exist, with different actors?
- Which actors can be eliminated?
- Are there actors who scale value streams multidimensionally or better?
- Is the business ecosystem robust and able to survive in the new scenario?

EXPLORE

BUILD/TEST

Describe the actors

- Who are the actors in the business ecosystem?
- What is their function and role in the system?
- What is their motivation to participate in the business ecosystem?

Prototype, test, and improve the business ecosystem

- With which MVE do we start?
- How and where can we test the MVE?
- What experiences help us to get the value streams, business models, and the role of actors in the ecosystem to improve iteratively?

Analysis of the advantages and disadvantages of each actor

- What are the advantages and disadvantages for each actor?
- What are his strengths/weaknesses and opportunities/risks in the system?

Multidimensional view of the business models

- What does the resultant business model and value proposition for each actor look like?
- How does the respective business model contribute to the core value proposition?
- Is the defined core value proposition the result of the sum of the value propositions of all actors?

EXPERT TIP
Factors of success for business ecosystem design

In order to apply a paradigm based on business ecosystem design successfully, you must keep five factors of success in mind:

1. Ecosystem awareness:
We should see ourselves as part of the ecosystem and develop the ability to recognize our roles and behaviors in it—also through the eyes of other people and actors as well as from various angles.

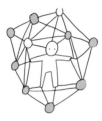

2. Understanding systemic options:
We should consciously reflect on ecosystems and have the ability to imagine what productive behaviors are possible for ourselves and for the whole ecosystem so as to change the value streams in a targeted way. We start with an MVE and extend it stepwise.

3. Managing of ecosystems:
Sharpen our ability to work with, on, and in the system, to integrate partners (co-creation) and create advantages for all actors.

4. Sustainable ecosystem intelligence:
Establish the capability of promoting and improving systems thinking and design thinking in this area in the long term and further developing the ecosystem in an agile way.

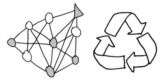

5. Leadership with business design ecosystems:
Build out the ability to integrate systems design into the culture of our organization and also to break existing rules consciously (black ocean).

KEY LEARNINGS
Design business ecosystems

- Reflect on the factors of success of systems thinking for the design of business ecosystems.
- Accept the complexity of ecosystems and always keep the big picture in mind.
- The customer's daily business and the customer needs constitute an important basis for many business models in an ecosystem-based approach.
- Put the user and the actors in relation to the value proposition, the complementary offerings, and the participants in the network.
- Connect the actors with the value streams, such as information, money, products, or digital assets and crypto-currencies.
- Also think about how the actors in the ecosystem will earn money and show them possible sources of income in order to make your business ecosystem appear attractive.
- Eliminate specific actors in the business ecosystem design already beforehand (e.g., intermediaries), who on the basis of technology leaps are no longer relevant.
- When developing the business ecosystem further, always focus on the customer experience and the growth of the platform and test the new functions quickly and iteratively.
- Use the business ecosystem canvas in order to document the (re)design and to follow the described procedure.
- Create a minimum viable ecosystem (MVE) and extend it stepwise.

Design thinking underwent various eras in the past. There was "synthesis" in the 1970s, followed by "real-world problems," all the way to business ecosystem design. Across all eras, we were faced with the challenge of successfully implementing the solutions in our organizations.

How do we overcome the hurdles in implementation?

We know from experience that various stakeholders in the company want to have their say. The process toward the solution is often scrutinized. The colleagues from the Legal department already raised objections to our very first prototype; the experts from Technology are generally not open to solutions they haven't developed themselves ("Not invented here" syndrome); and the hip ones from Marketing are bound by strict specifications for the branding of the new solution. In addition, there's the opinion of Management, the concerns of the Product Management Board, and all the other countless committees who question our ideas and block implementation. In most large enterprises, we will encounter resistance of a similar kind, not least because a rather traditional innovation and organization approach dominates most organizations. It is characterized by minimization of errors and maximization of productivity, the desire for reproducible processes, elimination of uncertainty and variation, and the need to increase efficiency with best practices and standard procedures.

NOT INVENTED HERE SYNDROME!

Design thinking as such offers a great basis for initiating transformation and innovating in an agile way. We consciously bank on interdisciplinary teams and radical collaboration, promote an experimental process through iterations, and thus maximize the learning success. Nonetheless, many of our ideas fall by the wayside and never get to market. As mentioned, one of the core problems here is that the stakeholders in the company are not part of the creative nucleus and often act within the structures and mindset of yesteryear. The style in which leadership is lived is frequently resistant to change in implementation projects. Even at a late stage—just before the market launch—willingness to change is greatly in demand. Just prior to the launch, we might realize that others were faster at bringing the solution to market. This is the point at which we ought to ask ourselves a number of important questions, because the answers to them will ultimately decide if our idea is on top or a flop:

- How can we nonetheless achieve a market success with other approaches?
- Have we thought through all types of business models?
- What value proposition creates a "market buzz" on the part of the customer?
- Are there any possibilities of partnership projects in the ecosystem so that our solution gets scaled?

As described in earlier chapters, collaboration with partners is becoming more and more vital for success in the digitized world. Many components can be provided by companies in the business ecosystem, in particular when it comes to technologies that we don't master ourselves. Collaboration can be stimulating for the development of the actual idea. The advantages are obvious: We boost speed and efficiency, participate in new trends and technologies, and reduce development costs. As a company, however, we need the ability to deal with open innovation models for this, especially in the form of collaboration, and explorative/exploitative skills. With regard to the latter, we are swiftly confronted with the problem of intellectual property (IP). But what seems to be even more important today are aspects such as who owns the data of digital solutions and the question of usability.

How can we, as an "oil tanker," act like a speedboat?

In the collaboration of traditional businesses with start-ups, two cultures clash, each of which follows different rules and has different hierarchies. Companies that are open to a so-called intrapreneur approach have the opportunity to develop a corporate culture in which higher risks are taken. Such approaches benefit from the fact that autonomous teams develop their ideas independently and with an entrepreneurial spirit, scale existing innovations, or establish themselves on the market as a spin-off. This approach is antithetical to the idea of central decision making etched in the DNA of traditional companies; in addition, it holds the risk that the skills provided and limited resources will not be able to scale the idea successfully on the market.

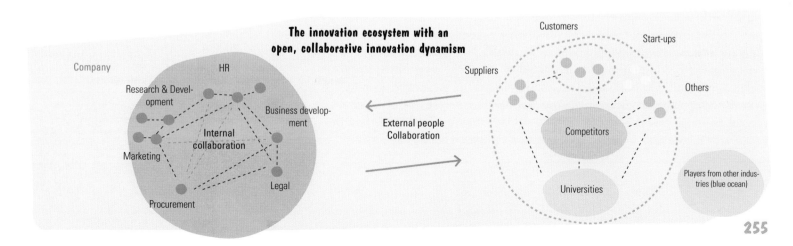

The innovation ecosystem with an open, collaborative innovation dynamism

How do the developed solutions get to the market?

For Lilly, her work at the university was usually over once the participants had presented their results. The implementation of the solution was not a major issue for her. Yet the feedback from industry partners and participants on the implementation made her think. Most celebrated opportunities do not find their way to market maturity.

In the case of rather simple market opportunities, the approach discussed before, namely to implement the idea with a start-up, is probably the simplest. You equip the teams with financial resources for one year, support them with coaching, and try to develop the opportunity to market maturity during this time. For one, this minimizes the risk; you also avoid all those hurdles that are simply written in the DNA of traditional enterprises. Such implementation variants have unfortunately been chosen only rarely up to now. This is because, among other things, it requires the willingness of the team to carry forward the idea in a venture for a while.

Execution is the key!

But Lilly also knows very well that implementation will be one of the most important factors of success in the future. From a study jointly conducted by HPI and Stanford in 2015, Lilly knows that design thinking brings forth many positive aspects in the field of work culture and collaboration. What is conspicuous, though, is that the top 10 effects of design thinking do not include an increased number of innovative services and products that were launched on the market. In our opinion, this is because implementation is not supplemented as the last and crucial phase in the process. As a mindset, design thinking must be established in the company holistically for it to be successfully implemented. Unfortunately, the reality is that the majority of responding organizations (72%) use design thinking in a more traditional way, namely in isolated areas of the "creative folks" in the enterprise—a fact that Jonny unfortunately observes at his employer as well. So we should not be too surprised that ideas are nipped in the bud of implementation.

IMPLEMENTATION =

We asked 200 people...

What are the effects of design thinking?

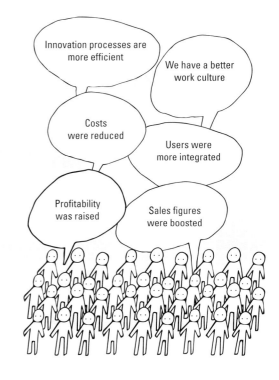

Create-ups are young enterprises that act as research laboratories. Behind these experimental labs, there is usually a strong entrepreneurial personality such as Marc and his co-founders. The founders often come from elite universities and bring in-depth knowledge from their studies as well as the capability of programming. Their intention is mainly to design new business models. They live a culture that is characterized by:

* a clear vision,
* a long-term strategy,
* a strong personal commitment on the part of the founder, and
* a high level of risk tolerance.

A typical approach for a create-up is to turn existing business models on their head and thus rock the market. They create a value proposition, for instance, that offers added value at better price/performance conditions or disrupts an existing business ecosystem. The employees in a create-up are passionate about the vision of the enterprise. Instead of power games, hierarchies, and rigid structures, the prevailing culture is one that aims at solving problems. Paired with a positive mindset, they are focused on realizing market opportunities. In so doing, they make intense use of the network that was already built up and cultivated at the elite universities. See Chapter 2.2, where we described the Connect 2 Value framework.

Established companies can learn a great deal from the create-up mindset. It starts with a clear vision, continues with a network-oriented way of thinking, and is expressed in the fact that all actors are involved in the problem solution as well as in agile and flat organizational structures.

CREATE-UPS

~~START-UPS~~

Traditional organizational models

No project teams

Experimental laboratories

EXPERT TIP
Involvement of all participants— working with stakeholder maps

Since we are often far removed from the create-up mindset in large organizations, it is all the more important to involve the relevant actors outside the company as well as the stakeholders within the company actively in the problem-solving process. The motto is: Turn the parties affected into participants. If we get the relevant stakeholders on board in the initial phases of the design process, they will understand much better why a problem statement has changed, which needs potential customers have, and what functions are important to a user. As soon as this understanding is given, all stakeholders will usually help proactively to bring the solution to the market. We will still sense resistance, but it will be far less than it would be were the stakeholders involved only late in the process. We have found the lean canvas approach to be a good way to do this (see Chapter 3.2). Within the company, Marketing can be included in the development of customer profiles, company strategists are closely involved in working out the financial aspects of the business model, and product managers see their responsibility in formulating an excellent value proposition. Thus all participants can identify more strongly with the development process when the solution is implemented later.

Just before the end of the design thinking cycle, it can be useful to take time again to draw up an implementation strategy. And it is even better to expend some thought on it a lot earlier. Especially in traditionally managed companies, a stakeholder map is very helpful for overcoming the many hurdles in such organizations in the best possible way. A stakeholder map helps to identify the most important actors and their relationships with one another. They are our internal customers to whom we need to sell the project. The key questions we have to ask ourselves:

- What are the current challenges from the CFO's point of view?
- How can the CMO best distinguish himself on account of our initiative?
- What does the product management board get from favoring our idea?
- How does the idea fit the grand vision of the CEO?
- How does the idea match the corporate strategy?
- Who is blocking the idea, and for what reasons?

The procedure for creating a real stakeholder map is simple. We take a good two dozen game pieces. Pieces that already have a certain character work very well from our experience. The pieces of Lego "Fabuland" are ideal. Then we need a large table that we cover with a sheet of paper. We put out Post-its and pens, a couple of ribbons, Lego bricks, and strings. The latter are used to create and visualize connections between the stakeholders. The discussion should be very open. In the end, the required measures are defined in order to approach the individual stakeholders in a targeted way.

Every large enterprise dreams of the flat and agile organizational structures of a create-up that allow for the quick piloting and implementing of market opportunities. Transforming an entire organization is a longer task, so a step-by-step approach is advisable.

Based on our observations, we recommend starting small and banking on a gradual transition. Ideally, we begin with a team that tries out agile working and experiments with it (first level of maturity). The focus is on learning how to work agilely. In a second step, we extend this approach to a second team that has characteristics similar to the first. These teams develop, for example, new functionalities for existing products in relatively short cycles, ultimately making customers happier with their ideas; or they experiment with a new business model for an existing solution. In a third step, you start to scale the agility to the entire organization. Multiple teams autonomously develop a complete business model, a product, or a service. A clear and unambiguous strategy helps the teams orient themselves and align their activities to corporate goals. It is important that these teams establish collaboration beyond the organizational units and that middle management is willing to cede responsibility. Agile program management constitutes the basis for a lean governance of projects. In a fourth step, we can replicate the approach further into the organization with the result of transforming the existing organization into an agile organization. The best indicator for the hoped-for goal is when the agile organization innovates from inside out. In the last step, the teams act autonomously, launch new initiatives within the set mission, and implement them "on the fly" (i.e., in short cycles) on the market.

This collaboration of various teams in a multi-program organization is also referred to as "teams of teams." "Guilds" or "tribes" are other terms frequently used in the context of such organizations.

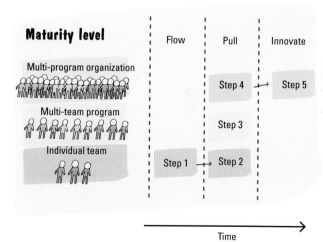

The work in squads is most similar to that in create-ups. One company that falls into this category is Spotify, whose management consistently relies on tribes, squads, chapters, and guilds. Like most create-ups, Spotify has a powerful vision ("Having music moments everywhere") that enables the individual tribes and squads to align their activities accordingly. Especially in technology-driven companies, a higher level of employee excellence can be achieved this way in a short time.

At Spotify, employees work in tribes. A tribe has up to 100 employees who are in charge of a shared portfolio of products or customer segments and are organized as simply as possible according to their dependencies. So-called squads form within the individual tribes that take care of one problem statement. Squads act autonomously and organize themselves. Experts from various disciplines are part of the squad and perform various tasks. Every squad has a clear mission; at Spotify, this can refer to the improvement of the payment or search functions, or of features such as Radio. The squads establish their own story and are responsible for the market launch. The mission is part of the clearly defined vision. The individual chapters ensure exchange on the technical level. Communities with the respective skills form, and are usually overseen by a line manager.

Guilds emerge on the basis of common interests. Interest groups form around a technology or market issue, for example, and then act transversally across the tribes. A guild might deal with blockchain technology and discuss its use in the music world of tomorrow. The organization is a network with flat hierarchies. The squads work directly with one another, and boundaries between them are fluid. In such structures, we tend to collaborate radically in order to solve a problem; for instance, we meet ad hoc and dissolve the collaboration in the same way. This way, networked organizations emerge. For such approaches, it is advisable to bid farewell to traditional role descriptions and hierarchy levels.

3 factors that lead to better performance and personal satisfaction:

- AUTONOMY
- PURPOSE AND MEANING
- PERSONAL RESPONSIBILITY

Character of autonomous squads:

- You feel like you are in a "mini start-up"
- Self-organization
- Cross-functional
- Five to seven people

When implementing market opportunities in large organizations, we cannot avoid measurability. The dimensions of traditional balanced scorecard approaches and well-known key performance indicators are not expedient and should be replaced with new questions. In particular, if the company is in a transformation phase, they must be reconsidered and/or discarded. We recommend including the key elements of a future-oriented organization in the cause-and-effect chain. The ability to think in ecosystems and the passion of teams for the execution of the mission can be crucial elements of management.

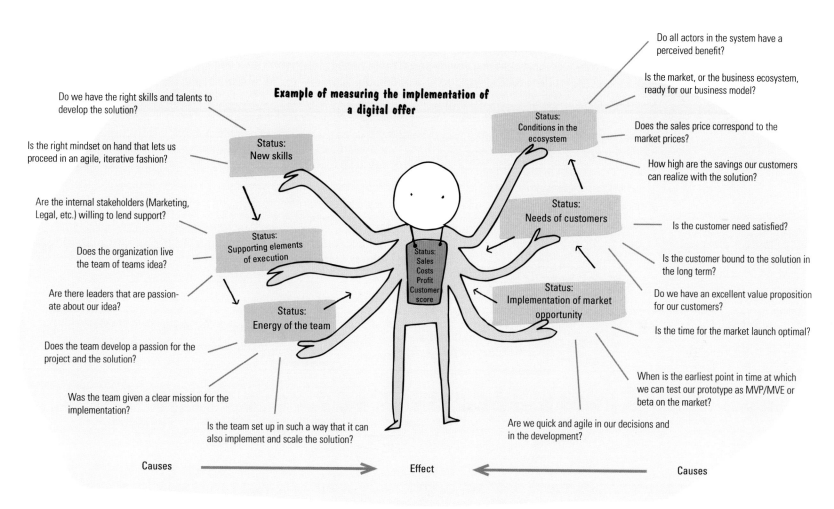

Example of measuring the implementation of a digital offer

Do we have the right skills and talents to develop the solution?

Is the right mindset on hand that lets us proceed in an agile, iterative fashion?

Are the internal stakeholders (Marketing, Legal, etc.) willing to lend support?

Does the organization live the team of teams idea?

Are there leaders that are passionate about our idea?

Does the team develop a passion for the project and the solution?

Was the team given a clear mission for the implementation?

Is the team set up in such a way that it can also implement and scale the solution?

Status: New skills

Status: Supporting elements of execution

Status: Energy of the team

Status: Sales Costs Profit Customer score

Status: Conditions in the ecosystem

Status: Needs of customers

Status: Implementation of market opportunity

Do all actors in the system have a perceived benefit?

Is the market, or the business ecosystem, ready for our business model?

Does the sales price correspond to the market prices?

How high are the savings our customers can realize with the solution?

Is the customer need satisfied?

Is the customer bound to the solution in the long term?

Do we have an excellent value proposition for our customers?

Is the time for the market launch optimal?

When is the earliest point in time at which we can test our prototype as MVP/MVE or beta on the market?

Are we quick and agile in our decisions and in the development?

Causes → Effect ← Causes

EXPERT TIP
Continuous exchange of ideas between the design teams

Innovation projects and problem-solving projects are used in different areas of the company. This obviously means that the time horizon and the definition of the future are also different for the individual design teams. In agile companies that are heavily based on technology, the time horizon for new services and products is usually no longer than one year. In the area of product groups, the cycles last between 12 and 24 months, depending on the industry focus. For weighty decisions regarding platforms with large investments, the standard is a time horizon of up to five years, not least due to the payback period. Strategic foresight as a design element has a perspective of five to 10 years. In addition to the desired market role, the question as to which business models will bring the revenue in the future is reflected upon. Furthermore, assessments are made as to how mega-trends affect the company and its portfolio. We have found that the continuous exchange of ideas transversally between the teams is a factor of success if we ultimately want to be innovative in a targeted way—always being aware, of course, of the time period the design team has in mind. In the terminology of a modern organization, the chapters make such a transversal exchange possible. In addition, the respective departments and business units, or squads and tribes, need the planning insights of an overriding strategy, so they can put their mission in the right context. Networking outwardly is a crucial factor of success.

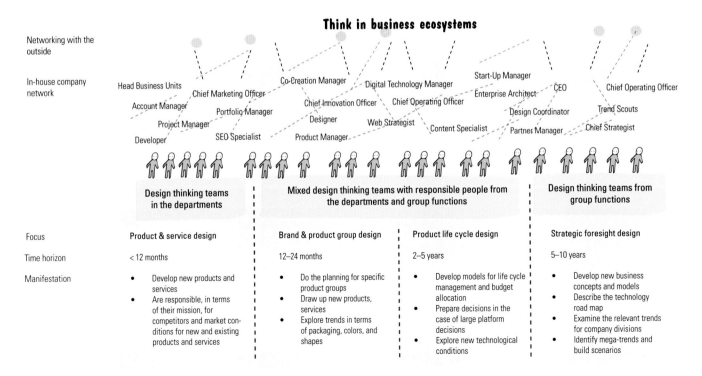

Think in business ecosystems

	Design thinking teams in the departments	Mixed design thinking teams with responsible people from the departments and group functions	Design thinking teams from group functions
Focus	Product & service design	Brand & product group design / Product life cycle design	Strategic foresight design
Time horizon	< 12 months	12–24 months / 2–5 years	5–10 years
Manifestation	• Develop new products and services • Are responsible, in terms of their mission, for competitors and market conditions for new and existing products and services	• Do the planning for specific product groups • Draw up new products, services • Explore trends in terms of packaging, colors, and shapes • Develop models for life cycle management and budget allocation • Prepare decisions in the case of large platform decisions • Explore new technological conditions	• Develop new business concepts and models • Describe the technology road map • Examine the relevant trends for company divisions • Identify mega-trends and build scenarios

KEY LEARNINGS
Implement solutions successfully

- Determine the relevant stakeholders in the company at an early stage and involve them in your design challenge.
- Develop an implementation strategy with specific measures on the basis of a stakeholder map, before the implementation is initiated.
- Establish agile and lean organizational structures that accelerate the go-to-market.
- Lend the implementation projects additional drive through external cooperation projects with partners, start-ups, and customers.
- Follow a step-by-step approach for the transformation into an agile organization. First, establish small and agile teams; then, scale the procedure with a clear strategy and guidance for the employees.
- Accept that not all projects, industries, and tasks are suitable for being realized in an agile organizational form.
- Always define a clear vision in agile organizations. Otherwise, tribes have a hard time specifying their tasks. Squads need the overriding vision to align their mission to it.
- Establish an awareness for the fact that the design teams have different planning cycles.
- Promote a transversal collaboration, such as through guilds.

Peter is fascinated more and more by the possibilities of digitization. Step by step, robots will be deployed on various levels and they will autonomously interact with us. Bill Gates once said: "A robot in every home by 2025." Peter believes that this development will take place even earlier. Cars drive autonomously on highways and private sites already, and new possibilities are continuously emerging in the area of cloud robotics and artificial intelligence. New technologies, such as blockchains, will allow us to carry out secure intelligent transactions in open and decentralized systems.

But what does that mean for the design criteria when we develop solutions for systems of tomorrow?

In the future, intelligent, autonomous objects will also be users and customers!

USE CASE AUTONOMOUS ROAD TRAFFIC

Vehicles will autonomously communicate, park, pick you up, drive.

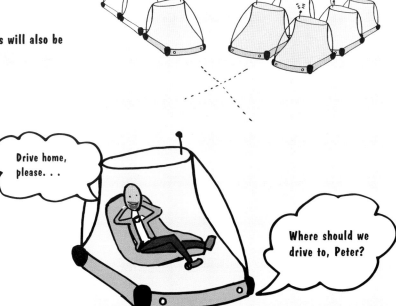

Drive home, please. . .

Where should we drive to, Peter?

In a nondigitized world, the relationship to people is primary for an improved experience. When we look at the development of digitization with its various priorities, the design criteria are extended over time. For the next big ideas in the field of robotics and digitization, new criteria become relevant, because the systems interact with each other and both (robots and human beings) gain experience and learn from each other. A relationship is created between the robot and the human being. They act as a team.

Therefore, among other things, trust and ethics become important design criteria in the human–machine team relationship. So-called cognitive computing aims at developing self-learning and self-acting robots with human features. Nowadays, many projects and design challenges, depending on the industry, are still in the transition phase from e-business to digital business. Digitization is thus a primary focus for companies if they want to stay competitive and exploit hitherto unknown sources of income through new business models.

TIME ⟶	1994 ⟶	2004 ⟶	2014 ⟶	FUTURE
	Analog/nondigitized world	Internet/Web	E-business Digital marketing	Digital business/ Internet of things	(Semi-)autonomous machines/ "robots"
Focus	Relationship with humans for a better experience	Expanded relationship in new markets and countries	Transformation of customer interaction into a global and efficient medium	Expansion of the relationship of people to machines	Intelligent, (semi-)autonomous machines interact with people and social systems
Design criteria	• Needs • Simple • Functionality	• Networking • Availability • Data	• Information • Business intelligence • Big data	• Knowledge • Prediction • Access to sensors	• Trust • Adaptability • Intention
Systems	• People	• People • Web	• People • Cloud	• People • Sensors • Objects	• People • Machines • Robots • Social systems • Cultures
Results	Optimized relationship	Expanded relationship	Optimized channels and interactions	New business models	Close relationship, human–machine as a team

What do the design criteria of the future look like?

The design criteria begin to change when the machines act semi-autonomously. In this case, human beings collaborate with robots. Robots perform individual tasks, while centralized control is still in the hands of human beings.

Things become really exciting when human beings interact with robots as a team. Such teams have far-reaching possibilities and can

- make faster decisions,
- evaluate many decisions synchronously while doing so,
- solve difficult tasks, and
- perform complex tasks.

Relevant criteria, which are to be fulfilled by a human–robot team, are inferred from the specific structure of a task. Design thinking tries to realize the ideal composition of task characteristics and characteristics of team members. But if human beings and robots will act together on a team in the future, the question arises of whether it is more important for us humans to retain decision-making authority or to be part of an efficient team. In the end, a good team performance is probably more important. Creating a functioning team is a complex affair, however, because three systems are relevant in the relationship between human beings and robots: the human being, the machine, and the social or cultural environment.

The great challenge is how the systems understand one another. Machines can simply process data and information. Human beings have the ability to recognize emotions and gear their activities accordingly, while both systems have difficulties in the area of knowledge. Knowing what the others know is pivotal! And then there is the element of the social systems. Human behavior differs widely due to its individual forms of existence in different cultures and different social systems. Not to be forgotten, ethics: How is a robot to decide in a borderline situation? Let's assume a self-driving truck gets into a borderline situation in which it must decide whether to swerve to the right or the left. A retired couple is standing to the right; on the left, there's a young mother with a baby buggy. What are the ethical values upon which a decision is made? Is the life of a mother with a small child worth more than that of the retirees?

A human being makes an intuitive decision in such a borderline situation, which is based on his own ethics and the rules known to him. He can decide himself whether he wants to break a rule in a borderline situation, such as failing to brake at a stop sign. A robot follows the rules it has been fed in this respect.

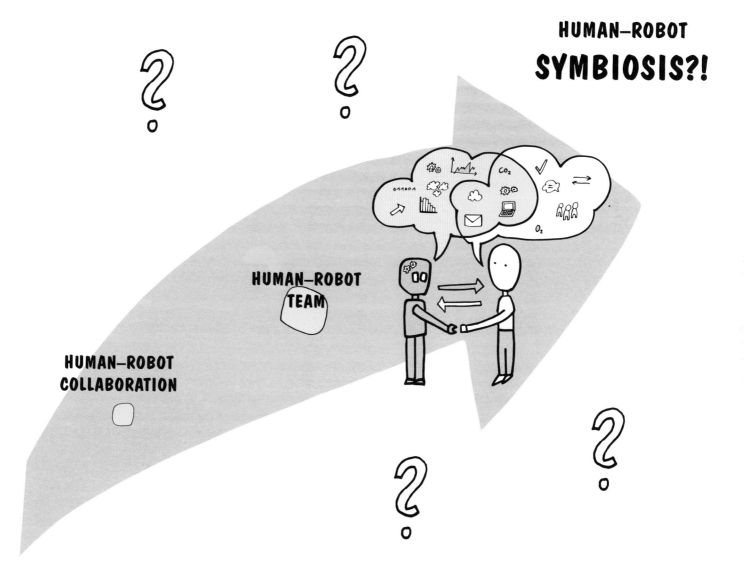

Even a simple action such as serving coffee shows that trust, adaptability, and intention in the human–robot relationship become a challenge for the design of such an interaction.

The question is: How should the world of robots and autonomous objects be integrated in the development of new digital solutions?

Today, Peter's design thinking reflections still focus on human beings. He builds solutions that improve the customer experience or automate existing processes. You might call it digitization 1.0. At higher maturity levels of digitization, things get far more challenging. With increasing maturity, robots also become more autonomous. Not only are individual functions or process chains automated, but robots interact with us on a situation-related basis. Thus they act multidimensionally. Trust, along with adaptability and intention, will be one of the most important design criteria. This means that good design will require all these design criteria in human–machine interaction in the future.

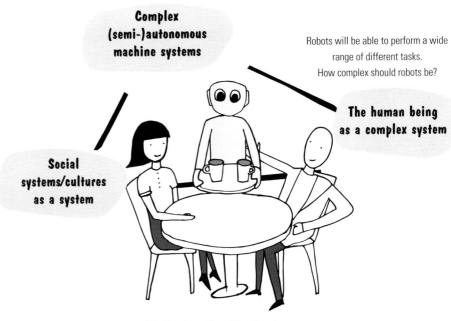

Who will be served coffee first?
"Ladies first"?
Should the coffee be served from the right or from the left?

Complex
(semi-)autonomous
machine systems

Robots will be able to perform a wide range of different tasks.
How complex should robots be?

The human being
as a complex system

Social
systems/cultures
as a system

Lilly likes her coffee with milk and sugar.
Peter wants his coffee without sugar but with soy milk.
We human beings are all different.

Peter has a new design challenge that he wants to solve in collaboration with a university in Switzerland. He's in contact with the university's teaching teams. Peter's design challenge comprises finding a solution for registering drones and determining their location. For the most part, today, autonomous drones are not yet out and about. But they are getting increasingly more autonomous and will fly by themselves in the future. They will perform tasks in the areas of monitoring, repair, and delivery; render corresponding services; or will be simply of use in the context of lifestyle applications.

Design challenge:
"How might we design the registration and tracking process of drones (> 30 kg/< 30 kg)/(> 66 lbs/< 66 lbs) on a central platform?"

The participants in the "design thinking camp" get down to work. A technical solution for registering the drones and identifying their location should be found quickly. Interviews with experts from flight monitoring corroborate the need for such solutions. An incident at a French airport when an airliner evaded a drone at the last minute during a landing only underscores this need.

Because all stakeholders are involved in such a design challenge, the students go one step further and interview passers-by in the city. They soon realize the general population is not very enthusiastic about drones and only accepts them to a limited extent. The design thinking team has come up against a much more formidable problem than the technical solution: the relationship between human and machine. Especially in the cultural environment of Switzerland, where the design challenge takes place, it seems important to pay heed to general norms and standards such as protection from encroachments on the part of government or other actors upon personal freedom. The participants see a complex problem statement here and reformulate their design challenge with the following question:

New design challenge:
"How might we design the experience of interaction between drones and humans?"

Based on this new design challenge, the question is illuminated from another side. The result is that the technical solution is put on the back burner, while the relationship between man and machine takes center stage in a more heightened way as the critical design criterion. Expanding the design criteria serves as a basis for a solution in which everybody can identify drones and, at the same time, can get expanded services from the interaction.

"I know who you are, and you seem to be friendly"

In this case, a prototype that was developed consists of an app that is networked with the potential cloud in which the traffic information on the drones flow together. Through the position data, the "Drone Radar App" detects the drone. The key feature is that the drone for which the information is retrieved greets the passer-by with a "friendly nod." This feature was quite well received by the people interviewed and shows how human behavior can minimize the fear of drones. Other prototypes also show that making contact in a friendly way or an associated service improves this relationship.

Because Peter has carried out the "drone project," he wonders where else robots will interact with humans in the future. What use cases are there?
Which senses can be captured by robots?

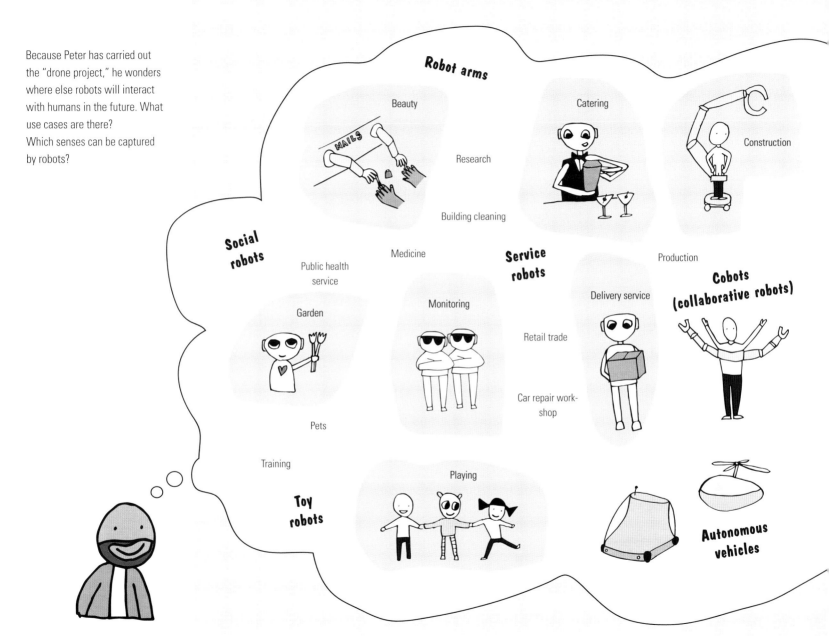

Robot arms

Beauty

NAILS

Research

Building cleaning

Catering

Construction

Social robots

Public health service

Medicine

Service robots

Production

Cobots (collaborative robots)

Garden

Monitoring

Delivery service

Retail trade

Car repair workshop

Pets

Training

Playing

Toy robots

Autonomous vehicles

EXPERT TIP
Coexistence of persona and robona

As the examples of autonomous vehicles and drones have shown, the future will be characterized by a coexistence between humans and machines. The relationship between humans and robots will be decisive for the experience. For initial considerations, creating a "robona" together with a persona has proven to be of use.

The creation of a robona arises from the human–robot team canvas (Lewrick and Leifer), with the core question being the one about the relationship between them. Interaction and experience between robona and persona are the crucial issues. For one, information is exchanged between the two. This exchange is relatively easy because certain actions are usually performed 1:1.

Things become more complex when emotions form an integral part of the interaction. Emotions must be interpreted and put in the right context. The exchange of knowledge requires learning systems. Only a sophisticated interplay between these components can properly assess intentions and meet expectations. That complex systems require complex solutions is especially applicable in this environment. The complexity is stepped up a notch in the human–robot relationship and its team goals.

Trust can be built up and developed in different forms. The simplest example is to give a robot a human appearance. Machines might come into being in the future that communicate with people and at the same time make a trustworthy impression on their human interlocutor. The projects of the "Human Centered Robotics Group," which has created a robot head that reminds the human interlocutor of a manga girl, are good examples of this. The creation, based on a schema of childlike characteristics (big eyes), makes an innocent impression—it makes use of the key stimuli of small children and young animals that emanate from their proportions (large head, small body). The robot also creates trust because it recognizes who is speaking to him: It builds eye contact, thus radiating mindfulness. Not only the way a robot should act but also what it should look like often depends on the cultural context. In Asia, robots are modeled more on human beings, while they are mechanical objects in Europe. The first American robot was a big tin man. The first Japanese robot was a big, fat, laughing Buddha.

Once robots become more similar to human beings, they can be used more flexibly: They help both in nursing care for the elderly and on construction sites. Trust is created when the robot behaves in a manner expected by the human being and in particular when the human feels safe due to this behavior. Robots that do not hurt people in their work—that stop in emergency situations—are trusted. This is the only way they can interact on a team with people. Both learn, establish trust, and are able to reduce disruptions in the process. The theme of trust gets more complicated in terms of human–robot activities in different social systems or when activities are supported by cloud robotics. Then the interface is not represented by big, trust-inducing eyes but by autonomous helpers that direct and guide us and thus provide us with a basis for decisions.

EXPERT TIP
Design of "emotions" with robots

Emotions in the robot–human relationship are just as important as trust. The human expects the robot to recognize emotions and to act accordingly. There is no doubt that human beings have emotions and that their behavior is influenced by emotions. Our behavior in street traffic is a good example: Our driving style is influenced by our emotions. And we react to the driving style of others. We are in a hurry because we must get to an appointment. We are relaxed because we are just starting off on our vacation. We drive aggressively because we have had a bad day. How does a self-driving car handle such emotions and tendencies? The robot must adapt its behavior and, for example, drive faster (more aggressively) or slower (more cautiously). If necessary, it must adjust the route because either we want to enjoy the scenery or get from A to B as quickly as possible. One possibility is that, in the future, we transfer our personality and our preferences like a personal DNA to distributed systems and thus provide a stock of information. Another possibility is for various sensors to transmit additional information in real time, which helps the robot make the right decision for the emotional situation easily and quickly.
Hence the recognition of emotions and the situational adaptation of behavior will be of an even greater significance in the future. The latest developments in this area include "Pepper," the humanoid robot of the Softbank telecommunications company that can interpret emotions.

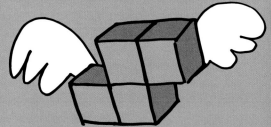

KEY LEARNINGS
Design criteria for a digitized world

- Accept that the customer of tomorrow might be a robot.
- Design interactions that reflect the coexistence between machines and human beings.
- Take advantage of the fact that humans and robots are most effective when they act as a team.
- Design all necessary areas of human–robot interaction. There is an exchange of information, knowledge, and emotions.
- Put the focus on trust. Trust evolves when the interlocutor behaves as you expect.
- Use a robona alongside the persona, so the interaction and the relationship can be visualized.
- Define a strategy that measures up to the dilemma of moral decisions being difficult for robots to learn and that they act in accordance with the algorithm that was programmed in them.
- Be aware that design criteria change and that complex systems require complex solutions.

3.6 How to kick-start digital transformation

Digital transformation is on everybody's lips, and the design thinking mindset, co-creation, and the radical collaboration of interdisciplinary teams on new solutions and customer experiences constitute a first step for initiating this transformation. Marc is already working in this mindset with his team. For him, an agile and iterative way of working is something obvious and part and parcel of the culture his team lives. In the preceding chapters, we have already shown some tools and methods for how even traditional companies can participate in this mindset. It was shown how we can

- define new paths with strategic foresight,
- (further) develop business models,
- realize new value streams by thinking in business ecosystems, and
- collaborate in an agile and networked way in new organizational forms, and so on.

There are many other categories that are equally important but, in most cases, very specific to the respective industry. It starts with our collecting and analyzing data and goes all the way to the maturity of automation or our willingness to drive forward decentralization and intelligence in open systems.

Starting digital transformation with a design thinking workshop

For product-centered and traditional companies, a design thinking workshop is frequently the initial leap across the digital divide and thus the kickoff to the transformation into a digital business. Digital leaders have a clear vision, have mastered the technology enablers, live a new mindset, and act with teams of teams when implementing their strategy.

For the various teams to be able to act on a self-organizing basis yet still with a distinct direction, a clear vision is required, from which the digital strategy can be derived. The direction in which the company should develop must be clear to all involved.

The new mindset will be crucial for the transition from a previously more deductive way of thinking to a design thinking mindset, the attitude of every individual and the entire team must match. For the interdisciplinary teams to achieve outstanding results, a positive energy is absolutely essential.

Every company and industry must define which technologies and technology enablers it needs and which will be a part of the value proposition. It's best if the knowledge of these key technologies be reliably available in the company itself or in the business ecosystem. One possibility here is the collaboration of start-ups or universities. Talent development with skills in the digital area must be ensured. Along with technical expertise, methodological knowledge and collaboration in interdisciplinary teams must be developed.

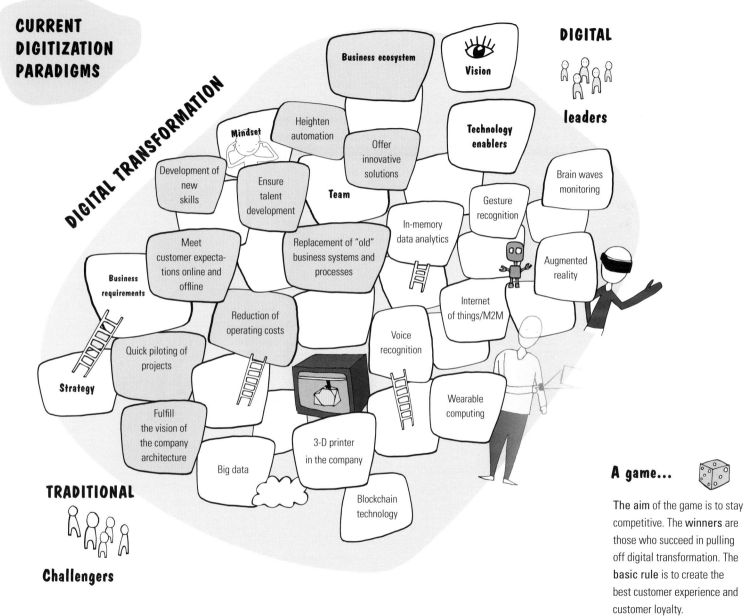

CURRENT DIGITIZATION PARADIGMS

DIGITAL TRANSFORMATION

DIGITAL

leaders

Business ecosystem

Vision

Mindset

Heighten automation

Technology enablers

Offer innovative solutions

Brain waves monitoring

Development of new skills

Ensure talent development

Team

Gesture recognition

Meet customer expectations online and offline

Replacement of "old" business systems and processes

In-memory data analytics

Augmented reality

Business requirements

Reduction of operating costs

Internet of things/M2M

Quick piloting of projects

Voice recognition

Strategy

Wearable computing

Fulfill the vision of the company architecture

3-D printer in the company

Big data

Blockchain technology

TRADITIONAL

Challengers

A game...

The **aim** of the game is to stay competitive. The **winners** are those who succeed in pulling off digital transformation. The **basic rule** is to create the best customer experience and customer loyalty.

279

HOW MIGHT WE...
initiate digital transformation and develop it step by step?

As mentioned, traditional companies must, metaphorically speaking, overcome a digital divide to bring about digital transformation. Previously valid assumptions no longer apply, including product-centered development, traditional hierarchical organizational structures, and a strong focus on market share and physical transaction chains with intermediaries. In many industries, this means a transformation of the whole organization. Possible steps to develop these skills might be:

1. Building up a new mindset with design thinking. New solutions and customer experiences are developed together with the customers (co-creation).

2. This type of work/collaboration is extended to the organization. As many teams as possible should be able to collaborate in agile and transverse fashion. Teams of teams are formed, and the organization changes from the inside out.

3. Many things can be scaled better if network effects are taken advantage of and integrated digital business ecosystems are developed instead of individual products and services with singular unique selling points.

4. This way of thinking can help, in a next step, to transfer the intelligence to decentralized structures and to implement business processes and transactions without intermediaries. Agile, transversal collaboration does not occur any longer solely on teams within the organization but across the whole company.

It takes courage to peer into the depth of the digital divide for the first time. The world of digital business is complex, and diverse, and it demands a new type of networked thinking on our part.

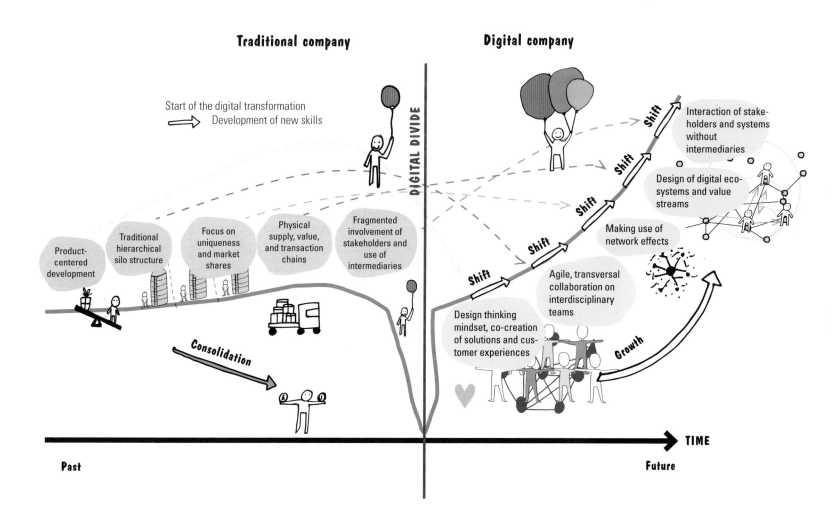

Traditional company

Digital company

Start of the digital transformation
Development of new skills

DIGITAL DIVIDE

Product-centered development

Traditional hierarchical silo structure

Focus on uniqueness and market shares

Physical supply, value, and transaction chains

Fragmented involvement of stakeholders and use of intermediaries

Consolidation

Shift

Shift

Shift

Shift

Shift

Design thinking mindset, co-creation of solutions and customer experiences

Agile, transversal collaboration on interdisciplinary teams

Making use of network effects

Design of digital eco-systems and value streams

Interaction of stakeholders and systems without intermediaries

Growth

TIME

Past

Future

First reflect and then transform with lots of positive energy

Too many times over the last few years have we had to witness how digital transformation was handled just like any other change management project: thought out, planned, and implemented from the top down. Measures were defined by the leadership team and taken into the organization by implying urgency. Unfortunately, none of these approaches were successful. For this reason, we should begin digital transformation with design thinking and then tackle it, as described, transversally across all company silos, involving everybody. At the end of the day, we want to have a system in which people work and have an effect. In our experience, it is good to give the people in the organization room so they can go through the cognitive process on their own and shape a new shared understanding in common—a mindset that matches the respective organization and its employees. In the "Business Ecosystem Design" chapter, in the context of the validation and realization loop, we pointed out the importance of the people in the respective companies, whose attitude and motivation are pivotal for the desired business ecosystem to develop positively.

However, the first necessary step is always to throw one's own assumptions overboard and exercise mindfulness. We can refer to this first phase as the phase of reflection. The following steps are in sync with our well-known design thinking principles.

The transformation is backed by many, not by a few chosen people. :-)

See technology as an opportunity for the change

It is also a fact that technology time and again provides us with opportunities for major upheavals. Digital and technological upheavals have already changed the world, and they will do so on a faster and more far-reaching basis in the future. We are at present in a phase in which blockchain, for example, as a technology enabler might herald in the next revolution. New market actors form up, and new value streams are defined. But it also means that industries that persist in their old patterns and individual intermediaries will be pushed from the market in the medium to long run.

What does this mean for traditional challengers?

Anybody wanting to survive in a blockchain world must have the skills and the mindset to engage in an ecosystem and shape this role actively. Most companies are still in a phase in which digital transformation is coming into being in randomly enacted and ad hoc initiatives, mainly driven by the automation of processes. We already mentioned that, for a higher level of maturity, large parts of the company should be involved in the exploration of new opportunities.

In individual cases, products and services are created this way that already have digital functionalities (e.g., sensors). With the steady dissemination of the design thinking mindset, a greater focus on customers and their needs will take hold. Digital solutions with a high level of customer focus make it possible to enter the market as a digital player, which is innovative on a transversel and networked basis. Opening up to the outside, including a heightened collaboration with partners, in order to set up innovative digital offers is decisive if you want to position yourself as a digital ecosystem player. From our experience, companies move on different S-curves when it comes to a digital orientation. Within the individual S-curves, the performance increases in the shape of an S. Combined forces and positive energy are needed between the respective orientations, however, to target the next S-curve of digital transformation.

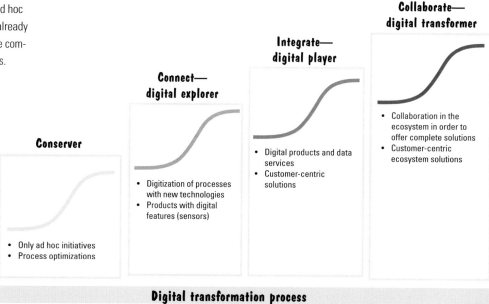

Collaborate— digital transformer
- Collaboration in the ecosystem in order to offer complete solutions
- Customer-centric ecosystem solutions

Integrate— digital player
- Digital products and data services
- Customer-centric solutions

Connect— digital explorer
- Digitization of processes with new technologies
- Products with digital features (sensors)

Conserver
- Only ad hoc initiatives
- Process optimizations

Digital transformation process

Why it is important to know the maturity of other actors in the business ecosystem

At the moment, Marc is pondering exactly these considerations about the actors in his business eco-system. He has analyzed how this transformation process runs on the S-curve for health insurers and discovered from which point on he will be able to establish a profitable value stream with insurers on the basis of his blockchain solution and data analytics on metadata. Marc has observed the following phases with respect to one health insurance company:

1st phase: Process optimization and cost savings as a result of digitization (e.g., automation)
2nd phase: Active penetration of multidimensional digital channels and digital processes; the digitization of process chains
3rd phase: Digital products and data-driven services are offered; the data generated this way is monetized and shared with other actors.
4th phase: Building a networked and trustworthy ecosystem in order to offer complete solutions that ensure market and scaling success

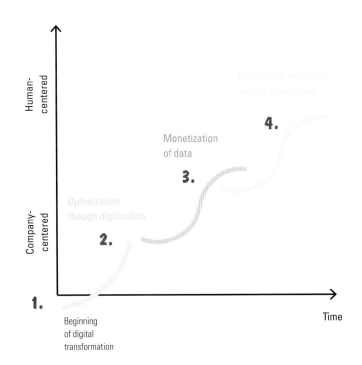

Sequence of a design thinking digitization workshop

We have pointed out that, for many companies, a design thinking workshop is a good way of initiating digital transformation. Of course, there are countless ways to bring the design thinking mindset to life. In our experience, specific questions work very well; other possibilities are to start with strategic foresight or deal initially with the business ecosystem design. What ultimately counts is in which industry the company operates and which "leaps across the digital divide" have already been made. We would like to present one example using Peter.

Peter also has the goal of exploring the potential of various companies with regard to blockchain. To do so, Peter has developed a two-day workshop that helps to get the opportunities of blockchain across to his industry partners; in addition, use options and business opportunities are discussed in the workshop. With his approach, Peter mainly addresses executives, and innovation and technology managers as well as other decision makers who are interested in the possibilities of blockchain and new models in business ecosystems. In addition, he invites actors and start-ups from the respective ecosystem so that a solution is developed in common right from the onset. Peter uses design thinking and the design thinking mindset for his workshop and combines them with business ecosystem design. At the end of the workshop, a decision is to be made on how the developed solution will be tested on the market.

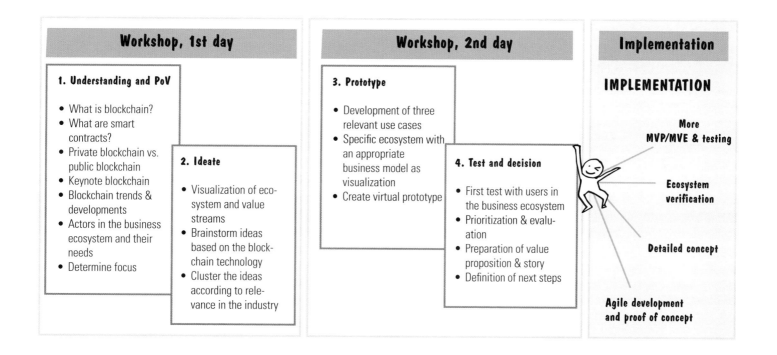

Workshop, 1st day

1. Understanding and PoV

- What is blockchain?
- What are smart contracts?
- Private blockchain vs. public blockchain
- Keynote blockchain
- Blockchain trends & developments
- Actors in the business ecosystem and their needs
- Determine focus

2. Ideate

- Visualization of eco-system and value streams
- Brainstorm ideas based on the block-chain technology
- Cluster the ideas according to rele-vance in the industry

Workshop, 2nd day

3. Prototype

- Development of three relevant use cases
- Specific ecosystem with an appropriate business model as visualization
- Create virtual prototype

4. Test and decision

- First test with users in the business ecosystem
- Prioritization & evaluation
- Preparation of value proposition & story
- Definition of next steps

Implementation

IMPLEMENTATION

More MVP/MVE & testing

Ecosystem verification

Detailed concept

Agile development and proof of concept

HOW MIGHT WE...
act if "digital" is not yet enshrined in the core of our business?

In a digital economy, the challenges are similar across all industries. The four most prominent are **dealing with uncertainty**, the **multidimensionality** of the business models, **participation** in business ecosystems, and the **scaling** (growth) that is necessary to achieve significant sales in these models. We have already shown possibilities for the definition of a forward-looking strategy. However, the question remains: What strategies can we apply in the short term if we have not yet managed to leap over the digital divide and the digital tsunami is already looming near.

The **first challenge** is uncertainty about how the future will develop. Uncertainty seems to rise in all industries. The good news is that "embrace ambiguity" is a crucial element in design thinking and its mindset can help us in dealing with it.

The **second challenge** is multidimensional business models. They usually serve several customer segments with completely different value propositions. We know such approaches in the form of multisided models such as Google. The customer usually "pays" with his digital traces in the network, such as with data he leaves in transactions and interactions. Other business models (e.g., freemium models) might be free of charge at the outset, and payment is only required for the use of additional offers. Such digital revenue models are somewhat complex; for many traditional companies, they are uncharted territory. In most cases, IT platforms, APIs, data analytics, and business ecosystems with the right partners for a target-oriented implementation are needed.

The **third challenge**, namely the handling and the design of ecosystems, was described in Chapter 3.3.

The **fourth challenge** consists of scaling the models and generating sustainable growth. Digitized business models are often not limited to national borders. Due to the shorter cycles, growth must be faster and have a broad basis. This also means that the infrastructure, structures, and processes must grow accordingly.

DIGITAL TSUNAMI

Traditional business

Many companies use one or more strategies to respond to these challenges, which we describe. It is advisable, though, to address the issue not only reactively and defensively but, instead, with a proactive stance.

(1) Blocking strategy

We attempt to prevent or slow down the disruption at all costs, such as by means of patent claims or the announcement of copyright infringements, putting up legal hurdles, and using other regulatory barriers.

(2) Milking strategy

We get the greatest possible value out of vulnerable business units prior to the unavoidable disruption (i.e., we milk the business as well as we can).

(3) Investment strategy

We invest actively in the threat. This includes investments in "disruptive" technologies, skills, digital processes, and possibly the purchase of companies with these attributes.

(4) Cannibalization strategy

We launch a new product or service that directly competes with the previous business model in order to build up inherent strengths ourselves for the new business, such as size, market knowledge, brand, access to capital, and relationships.

(5) Niche strategy

We concentrate on a profitable niche segment of the core market in which disruptions are less likely to occur (e.g., travel agencies with a focus on business trips or complex travel routes; booksellers and publishers focusing on the academic niche).

(6) Redefining the core strategy

We build up a new business model from scratch. It might even be in an adjacent sector if such a move makes optimal use of the existing knowledge and skills (e.g., IBM to consultancy, Fujifilm to cosmetics).

(7) Exit strategy

We get out of the business and return the capital to the investors. Ideally this is done by selling the company as long as it still has any value (e.g., sale of MySpace to News Corp.).

(8) Greenfield strategy

We start a new company in tandem with the old one on a greenfield site, which is then equipped with the necessary skills, infrastructures, and processes for digitization. We milk our old core business in order to build the new venture and switch as soon as we have successfully initiated the scaling.

What are the factors of success for the transformation of the business models?

If we opt for a redefinition of our current core strategy, for instance, it can be done on different levels. We can take a closer look at our specific value proposition and how it is generated, and optimize it. In so doing, we should always have an eye on the entire business ecosystem and our partner network because digitization facilitates new business models based on partnership or open business models. Alongside the constant screening of new technologies, heed must be paid to innovations and breakthroughs in the lower market segments (*jugaad* or frugal innovation). The drivers can be identified by the respective signals that are relevant to the industry in question (e.g., IoT and Industry 4.0); by trends in business models (e.g., shared economy); or technological revolutions (e.g., blockchains). The transformation of the employees is something not to be underestimated. We need new digital skills and must find ways for our workforce to acquire them.

What are the dimensions to be taken in due consideration?

When we want to design a new digital model for our company, four elements are absolutely critical to success from our experience alongside the focus on the needs of people: the operating model, the adapted technology, the digital business model per se, and, increasingly, the business ecosystem.
We can outsource a sub-function, for example, from our core business to an actor in the ecosystem in order to benefit from an expanded customer access. Or we already have established partnerships that we can profitably use for testing an MVE (minimum viable ecosystem). In addition, some thought should be expended on how complementary advantages through the combination of established platforms and digital services might be realized (e.g., a combined offering through a centralized touch point with the customer). In the definition of the model, it is vital to keep an eye on the big picture and at the same time explore it with minimal functions (MVP and MVE).

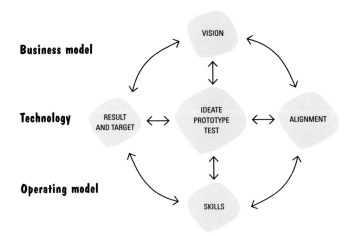

Business model

Technology

Operating model

VISION

RESULT AND TARGET ⟷ IDEATE PROTOTYPE TEST ⟷ ALIGNMENT

SKILLS

CONTENT
(information, product)

Product information
Price
Use cases
Digital products
...

DIGITAL BUSINESS MODEL
(experience)

Digitized business processes
Communities
Co-creation
Interfaces
Customer journeys
Touch points
...

PLATFORM
(internal, external)

Other business processes
Customer data
Technology
Proprietary hardware
Public networks
...

BUSINESS ECOSYSTEM DESIGN

Value streams
Win-win situation for all actors
MVEs

We have talked a great deal about fitness and, in the end, we might be able to cross the mountain pass with customer orientation, the right skills, motivated employees, and a good implementation plan. Companies that have higher ambitions and want to be among those who come in first in their own particular Race Across America should bank on "8 wins" (see p. 290). The Race Across America is one of the toughest bicycle races in the world. For such a race, it is not enough to tend toward aimless action in individual forms. What is required is the perfect match of material, willingness, and vision, all the way to the ability to integrate new technologies from an extended ecosystem. For our own organization, we must reflect on the question of where we are in terms of the individual forms. In the end, we must decide for ourselves whether we want to go on a leisurely ride on our bicycle or whether we have the willingness and the vision to compete with the best in the world. In 2015, it was David Haase who optimized his Race Across America using sensors, weather data, artificial intelligence, and big data/analytics. With this, David's performance reserves could be adapted to the conditions, and decisions were optimized.

Creating the Internet of Dave

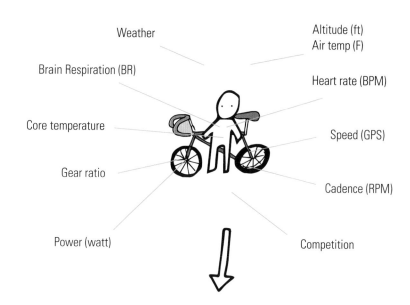

Weather

Altitude (ft)
Air temp (F)

Brain Respiration (BR)

Heart rate (BPM)

Core temperature

Speed (GPS)

Gear ratio

Cadence (RPM)

Power (watt)

Competition

The app/data
Analytics of Dave's app
Dashboard, predictive and race/rest decisions

The product
Real-time APIs
Forecasts for 25,000 geographical waypoints

The platform
IoT platform
The hub of the Internet of Dave; analytics accesses live data

Vision	Human–centered culture	Leadership	Teams of teams	The right talents	Positive energy	Ecosystem	Implementation and action plan	→	Hooray!
✓	✓	✓	✓	✓	✓	✓	✓	→	Successful transformation
✗	✓	✓	✓	✓	✓	✓	✓	→	Confusion
✓	✗	✓	✓	✓	✓	✓	✓	→	Market flop
✓	✓	✗	✓	✓	✓	✓	✓	→	No meaning
✓	✓	✓	✗	✓	✓	✓	✓	→	No agility
✓	✓	✓	✓	✗	✓	✓	✓	→	Frustration
✓	✓	✓	✓	✓	✗	✓	✓	→	Resistance
✓	✓	✓	✓	✓	✓	✗	✓	→	No effect
✓	✓	✓	✓	✓	✓	✓	✗	→	Stagnation

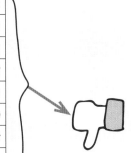

8 wins!

KEY LEARNINGS
Digital transformation

- Start the digital transformation with a design thinking workshop.
- Take the needs of customers into account when developing digital products and services.
- Accept the fact that new technologies will continue to bring forth great upheavals; at the same time, they give us the chance to tap new market opportunities.
- Overcome the "digital divide" by developing new skills (e.g., to use network effects).
- The greatest art in digital business models is to create business ecosystems and, as an entrepreneur, become a driver of digital transformation.
- Think in two directions when considering strategic options: either secure or milk the existing business and develop new digital business models.
- Digital transformation is also an organizational transition and demands agile and transversal collaboration on interdisciplinary teams.
- Establish a new mindset and teams in the organization, which can meet these challenges.

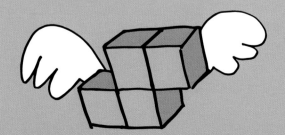

Designing a differentiating customer experience has become an integral part of the daily work in many customer-oriented companies. In Peter's company, a solid basis for the "next level of customer experience" has been already created as part of the digital transformation. People no longer think in departments but holistically and transversally. Along with Sales, Customer Service, Marketing, and Operations, partners and resellers get involved.

The primary focus at Peter's company is on differentiating the way the business interacts with its customers. Data plays an important part. After all, a multitude of it is collected with every interaction: from the visit in the brick-and-mortar shop to online purchasing all the way to interaction on the customer portal. A comprehensive customization can be carried out on this database for the individual customer. Because many companies today pursue a multichannel strategy, it is important to ensure that the individualization of the customer interaction is done across all channels and that each individual touch point makes a contribution to a differentiating customer experience. Thinking along the line of customer journeys is essential for the shaping of sustainable customer relationships.

I like that my customer journey is simple!

Simple and personalized interactions!

THE CUSTOMER EXPERIENCE ECOSYSTEM

Social media

Product

Mobile app

Call center

Web site

Environment

Physical

Shop

COMPANY
Brand
Employees

In the customer interaction, what are the challenges entailed in the digital customer life cycle?

In the past, the customer life cycle was more sequential and limited to a few channels—from perceive, inform, and order to install, use, and pay all the way to support and termination. It was usually done through only a few traditional channels with a (sometimes accepted) break in the media and the experience between the individual steps. As Peter has learned from a panel of experts at his alma mater in Munich, the customer of today moves through a broad variety of channels, sometimes concurrently; skips steps; and continues them in other channels, frequently with other devices. Digitization leads to new forms of interaction between the company and the customers (and between customers) and makes it possible to design more holistic experiences. These challenges and opportunities must be addressed in the design of customer experiences.

Identifying the customer concern and thus the phase of the customer life cycle in an interaction as early as possible constitutes one of the challenges. This is why it is essential for Peter's employer to collect the interaction data on in-house digital channels, classify the customer to the extent possible, and allow and use the data in the interaction.

The earlier the customer's reason for contact can be identified, the better the customer can be guided in the channel universe and the better the experience can be shaped. Assuming that the customer contacts the company in the event of a complex fault, the issue is not to prioritize the customer chat but to choose a channel that is more appropriate for the processing of the concern, such as video telephony or a technician visiting the customer at home.

With the increasing digitization of business processes, there are also more and more problems, which the customer can easily and without great effort process on his own on the customer portal. In addition, the company itself today has long ceased to be the only contact point for the customer along the entire customer life cycle. In fact, companies need to involve external value-creation partners. Customers can turn to communities, for instance, to have their problems solved. The Swiss IT company Swisscom, for instance, integrates both the online customer forum and the tech-savvy offline community of "Swisscom Friends" in the design of the experience.

When designing customer interactions in the digital life cycle, it continues to be crucial not only to allow for switching channels but to design the switch as an integrated experience. This includes making customer processes independent of channels, so a seamless switch is possible without information and status loss and without the customer having to state his problem more than once. The customer effect score or "easy score" can be used as an indicator here. These metrics show from the point of view of the customer how easy the company has made it for the customer to resolve his problem. Ideally, this indicator should be taken into account in the iterative development of customer experiences as early as at the beginning of customer tests.

TRADITIONAL CUSTOMER LIFE CYCLE

Discover
What solutions are there for my problem?

Ponder
What solution variants are there?

Use
How can I use it even better?

Buy
Which variant is best for me?

DIGITAL CUSTOMER LIFE CYCLE

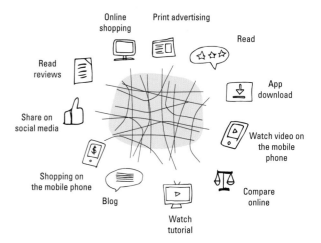

Online shopping

Print advertising

Read

Read reviews

App download

Share on social media

Watch video on the mobile phone

Shopping on the mobile phone

Compare online

Blog

Watch tutorial

How can we raise the service experience to a new level by means of the technological transformation?

In saturated markets, the customer experience is a major point of differentiation in terms of customer loyalty. The aim is for the customer to remember the interaction between the company and customers as a positive point of differentiation and develop a preference for the brand.

The technological transformation creates new opportunities to shape this differentiating experience in a more targeted way. Companies today have access to a huge amount of data that is created in interactions with customers, in processes, and with objects equipped with sensors.

Big data analytics makes it possible to process these large amounts of data and recognize patterns. The insights thus gained not only help us to understand the nature of the interaction between the company and the customer better and thus further develop the customer experience in general; they also enable us more and more to create an individual, differentiating service experience in the interaction with individual customers.

The developments in the area of machine learning allow us to penetrate areas in the service experience that were hitherto inaccessible on account of limiting factors such as finite human capital. Tasks previously performed by humans (e.g., in the customer dialog) can now (partially) be transferred to machines, which opens up the opportunity to scale new service models and implement them cost-efficiently.

The digital design thinking methods presented in this book help to design the "next-level service experience."

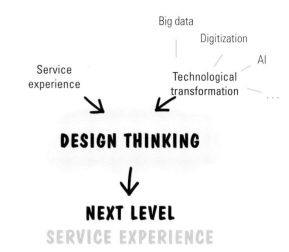

What new opportunities arise from using artificial intelligence in customer interaction?

With artificial intelligence (AI), we can finally open up the sweet spot of customer interaction: for many customers, a unique and personalized experience. In the past, an extraordinary service experience could be offered only to a selected group of customers. The great masses were denied this service experience on account of the high costs. All we could do was to present a rather limited service experience to the great masses, which then rarely left a lasting or differentiating impression. With the use of artificial intelligence, we are able today to create a personalized, high-quality service experience—and for a large number of customer groups (sweet spot), to boot. Hence service orientation can penetrate areas now that were economically not viable before. Digitization allows for the realization of so-called artificial assisted service models, which were too expensive to be performed by people up to now. Companies that know how to take advantage of this competitive technology advantage can become new leaders in the service area.

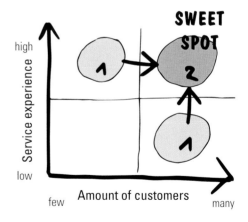

For which customer interactions should we rely on artificial intelligence?

AI tries to imitate human behavior when solving tasks. This implies that AI is capable of learning from its own observations and the existing data (e.g., interactions) to solve future tasks with the knowledge gained. One property of AI is that unstructured data such as text, language, and images—in other words, human communication—can be understood. In addition, the intelligence is increasing over time, because AI considers feedback from past decisions for future decisions. Compared to humans, AI not only decides more quickly and precisely, it is also able to take into account much larger amounts of context information. For us, this means we can transfer activities that follow a specific pattern to the machine when designing customer interactions. We deploy people only where specialized, nonroutine, and emotionally demanding tasks require it.

What might a possible vision of the customer dialog with artificial intelligence look like?

We recommend taking the first steps in working with AI in an area in which a great deal of interaction data is available that is recognizable and, hence, comprehensible to the machine. Within this area, applications can then be found within which routine activities are identified by means of AI and then outsourced to it in the future. This way, the benefit that is to be achieved by using AI in the customer interaction can be assessed quite clearly right at the onset.

Based on this initial experience, not only can more use options be better assessed, but more fields of application based on them can be found. One good starting point can be the customer dialog via e-mail, because this type of customer interaction still shows a great deal of potential for boosting efficiency and a solid database is usually on hand.

Example of a traditional interaction
- No answers
- Long waiting time
- Repeated interaction

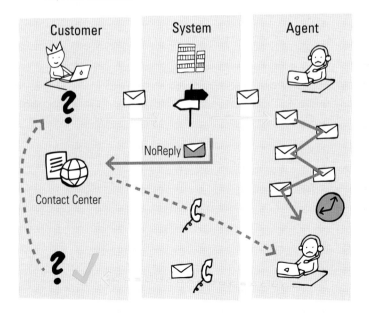

Example of support with cognitive computing
- Automated and appropriate response
- Faster processing
- Interaction adapted to customer concern

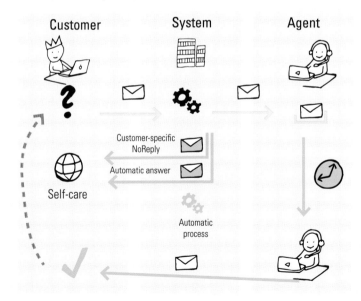

EXPERT TIP
Use of Social CRM

With Social CRM (SCRM), we can expand conventional customer relationship management using elements that focus on the interaction in social media. In so doing, we integrate in our CRM customer data that is not directly company related and use it as additional information in order to get a better idea of our customer and his interests. Through the collection of customer data via social media channels, we optimize our own service by acting on a more needs-oriented and proactive basis. For this, the willingness of the customer to share his data is relevant. It depends a great deal on how valuable he sees the sharing of his information with the company. It is crucial here that— from the customer's point of view—a fair exchange of (personal) information with the company takes place.

If, for example, we systematically collect the data traces of a customer's open Facebook account, we recognize the themes he discusses on social media. We can use this information as a trigger and thus approach him proactively with a matching offer.

An alternative to SCRM are so-called data providers, whose business model is based on the collection and sale of customer data. They sell information such as place of residence, shopping and travel habits, number of children and pets, clothing size, and so forth to interested companies. This information offers valuable insights for designing offers that match even better. It greatly depends on the ethics of each individual company to what extent such data is purchased and used.

The evolution from CRM to SCRM

CRM		SOCIAL CRM
Allocated to the departments	Who?	All
Company defines process	What?	Customer defines process
Business hours	When?	Customer determines the time
Defined channels	Where?	Customer-oriented dynamic channels
Transaction	Why?	Interaction
From the inside to the outside	How?	From the outside to the inside

EXPERT TIP
The Marketing Manager as digitization champion

The technological transformation confronts the Marketing Manager with a new challenge. He is successful if he puts the customer at center stage and, at the same time, is able to use appropriate technologies in the company. Today, the analysis of big data in real time—with the use of AI—is possible, as well as pattern recognition and prediction. A successful Marketing Manager makes use of this in order to understand his customers better and anticipate their needs. He performs data-based customer experience management so he can address his customers better and shape experiences actively. He can offer the customer a real added value. For example, he uses the movements or periods in life of his customers—so-called moments of truth—to create a perfect shopping experience or make the use of his product more relevant. In his function as Marketing Manager, he will become an innovator in the company alongside the traditional R&D and the Digital Manager. Many companies appoint additional, or evolutionary, Innovation Managers, who expedite an even closer integration of technologies and platforms.
A Marketing Manager as innovator will ask himself:

* Where does the customer stand in his period of life or even in his daily routine?
* What does the customer do, when does he do it, and where?
* What does the customer need at this moment?
* How can we reach our customers?
* Which data can I access?

These questions are intended to help the Marketing Manager enable a unique experience with the company for the customer.

Stakeholders

Innovate	Research & Development, Digital/Innovation Manager, Marketing Manager, CEO
Differentiate	Business units, Supply Chain, Process Manager
Start	IT Manager, Operation Manager, or Shared Services

The new role of the Marketing Manager

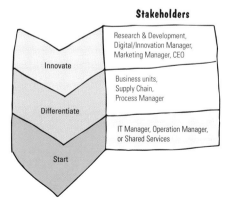

Applies hybrid models

Bases his decisions, among other things, on data

Focuses on the customer

Collaborates with partners in order to offer the customer multidimensional experiences

Plans to hire more employees with digital skills

Knows he will have more competition from companies outside his own industry

Offers digital solutions for customer loyalty

After discussing the role of the Marketing Manager as an innovator, we want to take a look at the role of the Innovation Manager, not least because it is also a role that's changing. The role of Innovation Manager is defined more and more by market requirements, changed business models, and the basic claim to understanding and using information technology (IT). In addition, the Innovation Manager has the tasks of bringing a new mindset into the organization on account of fundamental changes of market and business mechanisms, and of supporting the transition needed to adapt and to establish the right skills. He is also the link between the internal innovation systems and the external world, such as in partnerships with start-ups, accelerator programs, and universities. For this reason, it's a good thing if the Innovation Manager has built his network in the company over the years while working in other positions and if he is given the leeway to act freely with the external system. He performs his function as an inside outsider. In his function, he is innovative on the level of business models and below—always with a holistic view of emerging technologies and market needs.

The following questions are increasingly important to the Innovation Manager:

- Which pacemaker and key technologies do we need for new market opportunities?
- Which business models will be effective in our industry?
- Which start-ups and strategic alliances will bring added value?
- How do we increase the agility in the implementation of growth initiatives?
- How do innovation efforts become noticeable in the future scenarios?
- Which mindset suits us and how can it be spread transversally?

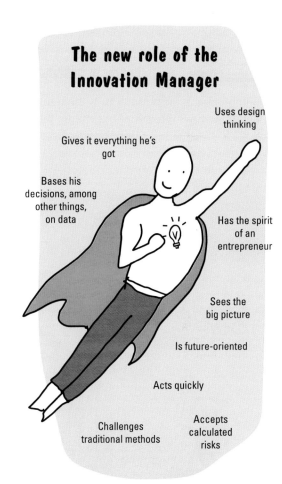

The new role of the Innovation Manager

Uses design thinking

Gives it everything he's got

Bases his decisions, among other things, on data

Has the spirit of an entrepreneur

Sees the big picture

Is future-oriented

Acts quickly

Challenges traditional methods

Accepts calculated risks

EXPERT TIP
The Digital Manager as an enabler of digital transformation

The Digital Manager is dedicated to the topics with top strategic priority in terms of the development of a digital offer. Today, he is usually the link between Marketing, Operations, IT, Innovation, and the Chief Executive Officer (CEO). The CEO has made the theme of digital transformation a core issue of the corporate strategy. The Digital Manager is tasked with providing the required skills, platforms, and technology components for a "seamless experience" and with implementing the digital initiatives. On account of a higher level of automation and maturity in the digital themes in many industries, Marketing will advance into an artificial inteligence department, hence changing the role of the Marketing Manager as a digitization champion. In many areas, the Digital Manager already takes on the tasks of customer interaction, communication, and their transition into digital experiences. In addition, he becomes an architect of digital ecosystems in which, on account of the new technologies, he also redefines the value streams and transforms the business models.

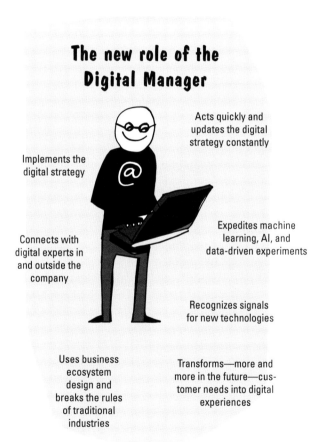

The new role of the Digital Manager

Implements the digital strategy

Acts quickly and updates the digital strategy constantly

Connects with digital experts in and outside the company

Expedites machine learning, AI, and data-driven experiments

Recognizes signals for new technologies

Uses business ecosystem design and breaks the rules of traditional industries

Transforms—more and more in the future—customer needs into digital experiences

KEY LEARNINGS
New technologies used for next-level service experience

- Use data, such as crowd-based data, for a better differentiation in the customer interaction.
- Think in customer experience chains and ensure that, in a multichannel strategy, the customer gets the best experience on every channel.
- Pay particular attention to the switch between channels and design such switches carefully so as to make the interaction as simple as possible for the customer.
- Use technologies such as artificial intelligence (AI) to realize next-level service experience.
- Create an affordable, very personalized, high-quality service experience for a great number of groups (sweet spot).
- Only rely on people in the interaction with customers when specialized, nonroutine, and emotionally demanding tasks must be performed.
- Use Social CRM to collect customer data in social media. Optimize with the data the service channels and act proactively and according to needs.
- Determine a digitization champion in the company. This can be a tech-savvy Marketing or Digitization Manager, who is an innovator and relies on the use of big data analyses in real time (with the help of AI), on pattern recognition, and prediction.
- Get the right skills and T-shaped employees into the company, who, for one, understand a technology and, secondly, can act and be innovative in ecosystems.

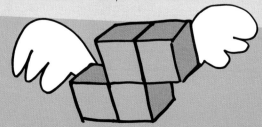

The job profiles and roles in our companies are changing across the board. There is a multitude of new job profiles today. Until recently, Peter thought he had the coolest job in his company. After all, as the Co-Creation and Innovation Manager, he shaped the innovations of tomorrow. Then, some time ago, he read in the *Harvard Business Review* that being a data scientist is the "sexiest job in the 21st century." In the future, data scientists will generate innovations, solve problems, satisfy customers, and get to know more about the customers' needs through big data analytics. In his blog on digital transformation, the CEO of Peter's company had also written about a data-driven business and that, nowadays, all business problems are solved with the new technologies.

How can we take advantage of this trend for our design challenges and integrate the faction of data scientists in the problem-solving process?

To benefit from big data analytics, we need a procedural model that combines design thinking with the tools of data scientists. The "hybrid model" (Lewrick and Link) is a suitable way to do so. This model has been developed based on the design thinking components. It promises to boost agility and ultimately result in better solutions. The hybrid approach gives companies the opportunity to position themselves as pioneers and become data-driven enterprises.

BUSINESS INTELLIGENCE

Better decision making because the decisions are based on data and not on intuition.

BIG DATA/ANALYTICS

Increasing amount of rapidly changing data. Usually collected by Internet companies.

HYBRID MODELS

Companies use big data analytics in combination with design thinking in order to improve their processes and range of offers.

- Design thinking
- Small + big data
- Analytics

DESIGN THINKING

The model consists of four components: (1) the hybrid mindset; (2) a tool box filled with the existing design thinking and new big data analytics tools; and other key elements are the collaboration of (3) data scientists with design thinkers as well as a hybrid process (4) that can give orientation to all parties involved. Thus the hybrid model is another possibility for expanding the design thinking mindset and generating better solutions from the combination.

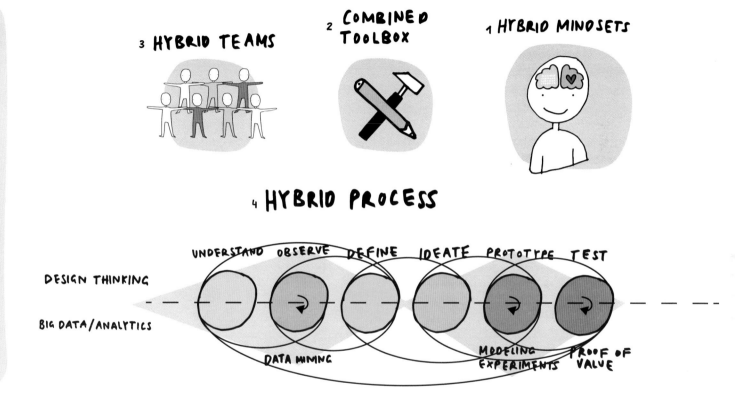

HYBRID MANAGEMENT MODEL

3 HYBRID TEAMS

2 COMBINED TOOLBOX

1 HYBRID MINDSETS

4 HYBRID PROCESS

DESIGN THINKING

BIG DATA/ANALYTICS

UNDERSTAND OBSERVE DEFINE IDEATE PROTOTYPE TEST

DATA MINING

MODELING EXPERIMENTS

PROOF OF VALUE

The advantage of the hybrid model: We create a mindset that gives us superior arguments when dealing with skeptics in traditional companies. One frequent point of criticism is that design thinking generates information on the needs only through ethnographic and sociological methods such as observation and surveys. With the hybrid approach, we can eliminate this vulnerability. Expanded by tools for the collection and analysis of big data, the quality of the design thinking process is heightened throughout.

HOW MIGHT WE...
go through the phases in the hybrid model?

Because the hybrid model is following the design thinking process, we primarily want to point out what is added. As in design thinking, the customer need and a problem statement (pain) to be solved mark the beginning. It can be a more rational or else an emotional problem. In the end, the solution may be a newly defined physical product, a digital solution in the form of a dashboard, or a combined solution that encompasses both elements.

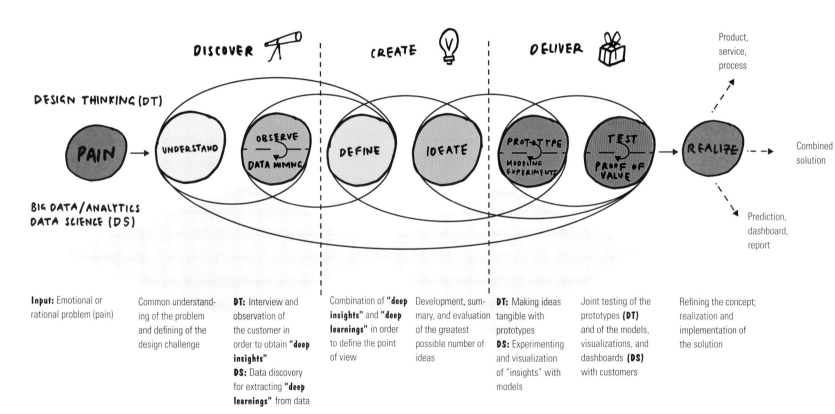

Input: Emotional or rational problem (pain)

Common understanding of the problem and defining of the design challenge

DT: Interview and observation of the customer in order to obtain **"deep insights"**
DS: Data discovery for extracting **"deep learnings"** from data

Combination of **"deep insights"** and **"deep learnings"** in order to define the point of view

Development, summary, and evaluation of the greatest possible number of ideas

DT: Making ideas tangible with prototypes
DS: Experimenting and visualization of "insights" with models

Joint testing of the prototypes **(DT)** and of the models, visualizations, and dashboards **(DS)** with customers

Refining the concept; realization and implementation of the solution

1 The first phase is **understand:** We develop in common an understanding of the problem. It is important that data scientists and design thinkers already collaborate here. Some facts can be determined through the analysis of social media data, for instance, which has a broader base than data gathered from traditional user surveys.

2 The **observe & data mining** phase is dedicated to the collection of "deep insights" and "deep learnings." "Deep insights" arise from our traditional observations of customers, users, extreme users, and the like. To obtain "deep learnings," data must be collected, described, and analyzed, which allows us to identify initial patterns and visualize them. We recommend discussing the insights from both observations together and reviewing the next steps.

3 In the **define** phase, we combine the "deep insights" and "deep learnings." A more exact point of view can be defined this way. The PoV describes the need a specific customer has and on what insights the need is based. The combination of both sides helps to get a better picture of the customer. The stumbling block again here is the definition of the PoV. We already talked about it in Chapter 1.6. The hybrid approach yields more "insights" that confirm the PoV but can also result in even bigger contradictions.

4 The aim of the **ideate** phase is to continue to generate as many ideas as possible, which are then summarized and evaluated by us. Several ideas are available at the end of this phase that are used in the next steps.

5 Then comes the **prototype & modeling experiments** phase. In this phase, we develop prototypes and carry out experiments with models. Prototypes make ideas palpable and easy to understand. As we know, a prototype can take many different forms; an algorithm, for instance, is also a simple prototype. The insights from the data experiments are best represented with models in the form of visualizations; in data science, this is the best solution to make something tangible.

6 In a **test & proof of value** phase, the prototypes are tested together with the potential user in order to learn from the feedback and adapt the solutions to the needs of the customer. This includes models, visualizations, and dashboards from data science, which constitute the basis for the prototype.

7 In the final phase, **realize**, we transfer an idea into an innovation! This includes integrating the models in operations. While data solutions usually evolve from data science projects and design thinking develops products or services, in the hybrid process, combined solutions from data science and design thinking can emerge. This can refer to a service-plus business model that presents added value as a result of the aggregation of various data sources; an example would be changes in the behavior of drivers to avoid traffic congestion in combination with an app.

EXPERT TIP
Live a hybrid mindset

For the successful combination of big data analytics and design thinking, a mindset should prevail that reflects the work in a hybrid model.
Because we now have a group of data scientists on board in the projects, it is useful to add the corresponding components to design thinking. A possible mindset can be described as follows:

We combine human insights and data insights (e.g., in a PoV).

We accept uncertainty and interpret statistical correlations in the context of the user.

We live an optimistic approach and iterate across the entire design cycle in both thought preferences.

We reflect on our approach and develop the hybrid mindset constantly.

We combine analytical and intuitive ways of thinking in a holistic approach.

MINDSET OF A HYBRID THINKER

We are interested in the unknown and create clarity through a creative and analytical approach.

We welcome the combination of design thinkers and data scientists. Mutual inspiration becomes a factor of success.

We generate stories from data and experiences in the form of prototypes and visualizations.

We learn to handle creative and analytical tools by trial and error and the acceptance of mistakes.

You need an interdisciplinary team to work with the hybrid model. It is made up of design thinkers, data scientists, and those responsible for implementation.

A facilitator who has the methodological knowledge continues to support the team. The team members can come from a wide variety of areas and contribute their differentiated background knowledge. Depending on the situation, the right specialist from data science can be used. The people responsible for implementation are part of the team.

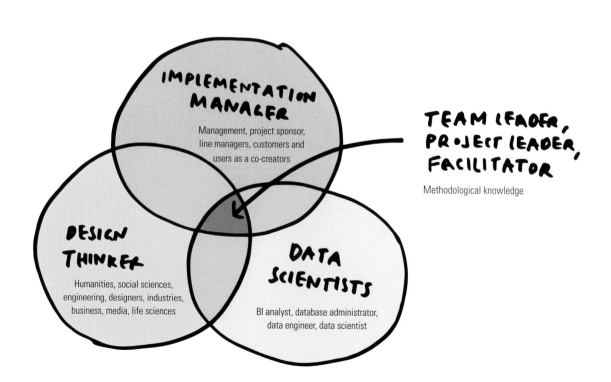

IMPLEMENTATION MANAGER
Management, project sponsor, line managers, customers and users as a co-creators

TEAM LEADER, PROJECT LEADER, FACILITATOR
Methodological knowledge

DESIGN THINKER
Humanities, social sciences, engineering, designers, industries, business, media, life sciences

DATA SCIENTISTS
BI analyst, database administrator, data engineer, data scientist

We recommend having a combined toolbox ready that contains the usual methods of design thinking and the tools from data science. As in design thinking, the critical point is: use the right method at the right point in time. There are many useful methods in design thinking that are easy to apply and quick to learn for everybody. In data science, things are a little more complicated because many tools require expert knowledge. But there is hope that more and more tools are being established that are user-friendly and can be used to perform data analysis without programming skills and expert knowledge. In addition, an increasing number of companies train their employees to acquire these skills. We have had very good experiences with Tableau. This is an easy-to-use tool. It also has a "back" functionality, if something goes wrong in the data experiments.

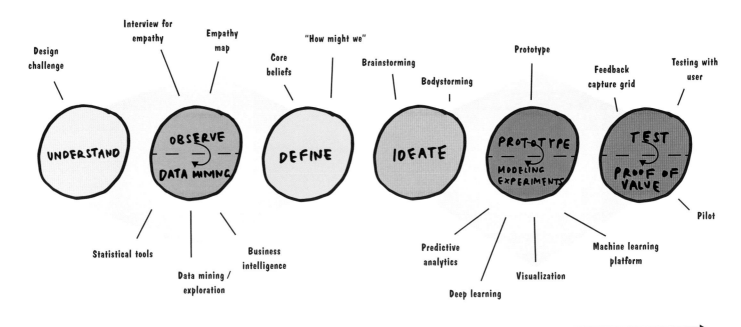

The hybrid approach compensates for the weaknesses of the unified approaches. Introducing a combined mindset has better chances of success than introducing one after the other sequentially.
In our experience, both top-down and bottom-up work.

With a bottom-up approach, the exchange between employees who deal with the subject of design thinking and those who are into data is promoted. In workshops, the two groups can present their approaches and challenges to each other. It quickly becomes apparent that the two approaches are complementary. The goal is to find a common pilot project in which the collaboration can initially be tested.
In a top-down approach, the advantages and disadvantages of both mindsets are presented to top management with the goal of carrying out an initial pilot project using the method of the hybrid model. After the pilot project is completed, the experienced gathered and the advantages are reported to top management and the stakeholders. In general, the hybrid approach reduces a number of risk factors; for example, it lowers the innovation risk of early experiments. In interdisciplinary teams, not only are new skills brought to the projects but also different ideas, which broaden the perspective. The same applies to a combination of systems thinking and design thinking and to projects that link strategic foresight to design thinking.

The hybrid approach—paradigm shift reduces risks

Paradigm shift	Risk factors that can be reduced
Focus on the overall picture (human being + data)	Innovation risk/risk entailed in search field for ideas
New mindset	Cultural risk
New composition of the teams	Skills risk
New hybrid process	Model risk

Implementation principles	Risk factors that can be reduced
Support, top management	Implementation risk
Part of the transformation toward digitization and/or data-driven enterprise	Strategy fit risk/management risk

EXPERT TIP
The supreme discipline—varying mindsets in the double diamond

The usefulness of hybrid models became quite clear to us early on. Along with heightened agility, we can generate more insights with the combined approach and mindset, which allows us to increase the number of possible solutions. Top innovators go one step further with their mindset and switch between design thinking, systems thinking, and data analytics across the entire development cycle. The quadruple diamond ensures that the optimum mindset is applied at each point in the cycle. Especially with far-reaching and complex problem statements, the respective design teams, squads, or experimental labs can optimize their work and apply the different skills sequentially or in a mixed form. The respective experts come from the corresponding chapters or guilds and help ensure that the necessary skills are available for each phase. As a facilitator or as the leader of a tribe, this also means having a higher level of methodological expertise at their disposal and having a sense for applying the right methods and tools in each phase.

We benefit from the fact that the three approaches go through similar steps. Thus the quadruple is purposefully built on the "double diamond" of design thinking, which is augmented by data analytics and systems thinking. Depending on the project, the mindsets can be mixed in the respective iterations.

When applying them sequentially, one single approach is executed; in the reflection at the end of the iteration, the further course and the method to be used in the next iteration are determined.

As described in Chapter 3.1, using design thinking and systems thinking in every project is recommended.

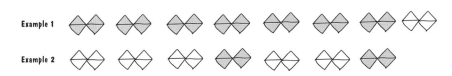

In a project that is largely driven by design thinking (example 1), systems thinking should be applied at least once in the end so as to depict and classify all the insights systematically. In a project that is driven by systems thinking and in which the system has already been improved iteratively two or three times, the critical assumptions should be checked in design thinking experiments, thus validating the system (example 2).

At the end of the day, the point is to understand each and every aspect of the problem from all perspectives by working on mixed teams with mixed methods. In the second part of the double diamond, the right solution is then also found with combined approaches. Example 3 shows the combination of design thinking and data analytics.

You can also combine all three approaches. This should only be done by experienced teams together with a facilitator, however. Combining all approaches naturally requires know-how in all of them (see example 4). As in the hybrid model, it is important that mindset, team, and tool sets are combined and not only the process be considered.

QUADRUPLE DIAMOND

Stronger than a triple

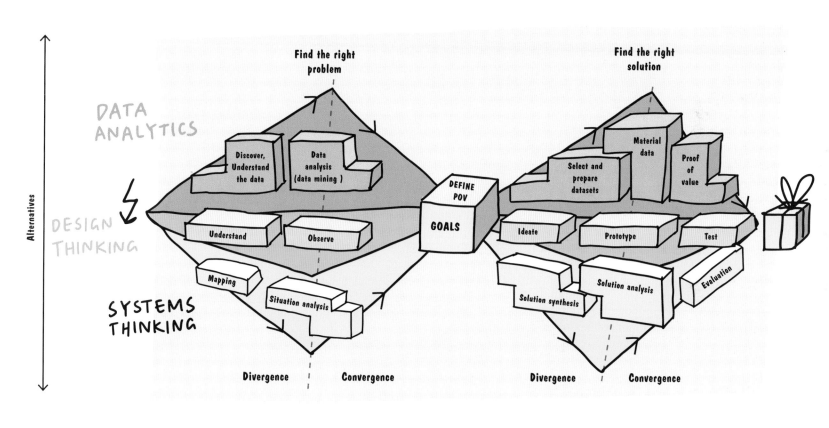

As in the hybrid model, it is important that mindset, team, and tool sets are combined and that not only the process be considered.

QUADRUPLE DIAMOND

Stronger than a triple!

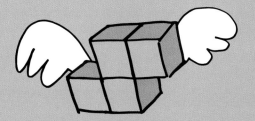

KEY LEARNINGS
Hybrid models—combination of data analytics and design thinking

- Take advantage of the technological possibilities of big data analytics and machine learning in order to be innovative in a more agile way.
- Define a problem statement in conjunction with the new contingent of data scientists.
- Observe people and data and draw conclusions in common.
- Accept that deep learnings can also come from data.
- Establish a common design process and make clear what should be done in what phase.
- Develop a common mindset to establish a hybrid model successfully.
- Work on mixed teams. Take data analytics experts on the team.
- Accept solutions and prototypes based on data.
- Be open to innovations that are more data-centric than human-centric.
- Win over top management for a hybrid model or start bottom-up with a pilot project.
- Use alternating mindsets and combine design thinking, systems thinking, and data analytics as the supreme discipline.

Closing words

There are quite a few things we have learned on the journey through the *Playbook* and in the interaction with potential users, our readers. The time has come to reflect on the factors of success before we move on and say goodbye to Lilly, Peter, Marc, Priya, Jonny, and Linda.

We have been confirmed in our claim that the traditional management paradigms must be challenged in order to detect future market opportunities and successfully implement them. The traditional mechanical-deductive approaches will make it difficult for companies to redefine entire value-creation chains and adapt their business models to the new customer requirements. Yet, unfortunately, the idea that innovation follows a defined stage-gate process with a clear sequence from search for ideas through implementation is still prevalent in many companies and in the minds of directors, department heads, and those responsible for innovation. These models have been outdated for at least a decade now.

There is a need for a fundamental change toward a systemic-evolutionary approach. Ideally, highly motivated interdisciplinary teams act in self-organizing network structures. Their work is based on customer needs and is geared to implementing new services, products, business models, and business ecosystems in a targeted way. When searching for the next big market opportunity, design thinking offers a strong mindset. This mindset must be further developed and combined with other approaches. No one size fits everything—we have to find our own way and the appropriate mindset for our organization.

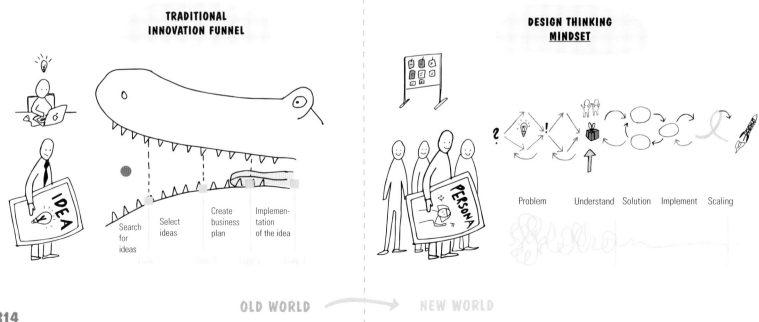

TRADITIONAL
INNOVATION FUNNEL

Search for ideas · Select ideas · Create business plan · Implementation of the idea

DESIGN THINKING
MINDSET

Problem · Understand · Solution · Implement · Scaling

OLD WORLD → NEW WORLD

After interacting with and observing numerous employees in management positions of companies that already have initiated the transition, we noticed in the context of innovation, co-creation, inspiring teams, facilitation, and the like, that design thinking still has not been comprehensively integrated in most companies. Design thinking initiatives often run in specially dedicated organizational units, and a new mindset prevails sometimes more strongly, sometimes more weakly in the companies. Likewise, we could recognize little activity so far that aims toward a consistent application of a hybrid model, with the combination of big data analytics and design thinking. Agility could be boosted and the number of possible solutions increased with such a combination. The business ecosystem approach has so far been dominated by only a few players. Most product managers in large companies have so far neither learned the necessary skills nor received instructions to monetize new solutions in decentralized networks. In addition, a clear vision is lacking in many companies and the same goes for the use of strategic foresight or systems thinking, for instance, in combination with design thinking for an improved mapping of a possible future. Change takes time as well as strong personalities with a clear vision at the top of companies.

However, we hope that a mindset can also spread on a bottom-up and transversal basis in an organization. When we look after our day-to-day tasks tomorrow morning, we should try to live the new mindset or think about which mindset might fit our organization. So heads up and get going!

A company without a strong vision of the future will always return to its past.

Companies with a clear vision of the future give their teams meaning and purpose and show what market role they would like to take on in the future.

OLD WORLD → NEW WORLD

BUSINESS THINKING :
HEADS - DOWN

Focused on deadlines

DESIGN THINKING :
HEADS - UP

Focused on opportunities

We have noticed that the training and the application of design thinking at universities and colleges are spreading very fast. Problem-based learning is an important experience for many participants and students not only in a design thinking course—it's a contemporary way of conveying learning content in general! They learn how agilely and radically teams work and network with other participants in the course of actively processing the problem statements. Often, this gives rise to connections that help to find suitable co-founders for the implementation of business ideas. Companies should also become more involved and have their problem statements prepared by the students. Such possibilities are offered by the ME310 course at Stanford University: in the framework of SUGAR, various international student teams are searching for the next great market opportunity. In addition, well-known programs are offered in Europe and Asia plus at numerous institutions in North America.

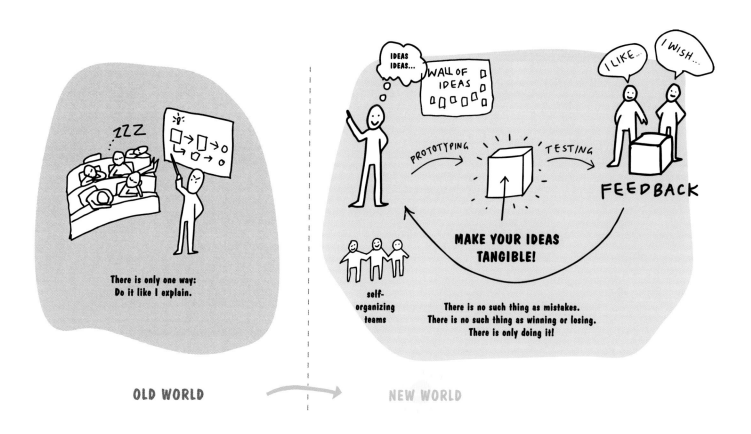

OLD WORLD → NEW WORLD

Tschüss, Peter! Zàijiàn, Lilly! Bye-bye, Marc! Hello, Future!

Peter, Lilly, and Marc have accompanied us as personas through the *Playbook*. And we hope that some of you have recognized yourselves in the day-to-day challenges, problem statements, and character traits. We have grown fond of the trio!

Lilly has now fleshed out in detail her design thinking consulting services and has dared to found the company. The tagline of her consulting boutique is: "The future is not enough." Jonny has sent her way an initial project at the bank for which he works. This customer is an excellent reference for Lilly's value proposition. For this proposition, Lilly relies on the soft approach of design thinking and the lean start-up methodology. In addition, she integrates a number of models from the old world in her own design thinking approach—"Shaken, not stirred"—something that is quite well received at least by global banks. After all, some bankers drive an Aston Martin like James Bond in the old days. A proposition that would better incorporate Asian ways of conducting business is unfortunately still lacking. Moreover, Lilly is three months pregnant and looks forward to little James completing her happiness in the near future. By the way, Lilly didn't tell us in the beginning that she is such a big Bond fan!

Peter is increasingly focusing on design challenges, which are on the agenda as part of the digitization efforts of his employer and his customers. He especially likes the new smart topics—from smart mobility to smart city. He dreams of a multimodal mobility platform for Europe. On this platform, all public and private mobility providers would offer their services. The people in the European Union and their visitors would have a unique customer experience to plan their trips optimally, book flights and train rides through state-of-the-art technology (e.g., blockchains), and process payment transactions. In addition, unique movement and behavior data will be made available to individual countries and cities in order to make everything a little bit smarter. Thus not only traffic would be directed optimally but also the planning of the urban space. This would be a first step toward a "smart nation" that is second to none! Peter is very well aware of the fact that this design challenge is a wicked problem, but maybe the challenge might be solved by design thinking in combination with systems thinking, the hybrid model, and, of course, the right mindset.

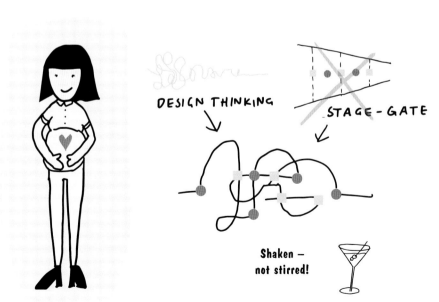

DESIGN THINKING

STAGE - GATE

Shaken —
not stirred!

Marc and his start-up team continue to think big. They want to revolutionize the health care system! The team realized the first functionalities of their idea for a private blockchain for patients. The current MVP has only limited functionalities but they are used in the MVE and it already provides the patients with information about which data has been created for the "health record" and when and where it was done. In the meantime, the start-up also cooperates with an established technology company that has shown interest in taking on the development for the analytics in the field of health metadata. The cooperation partners use the hybrid model for their innovation projects. Thus sound solutions emerge iteratively with the perspectives of the business team and data analytics teams.

Within the framework of the Start-X at Stanford University, Marc and his team also had the opportunity recently of pitching their business idea and the first MVP/MVE of their solution. At the end of it, several companies wanted a personal meeting with the team. In particular, the multidimensional business model in the business ecosystem, in which the benefits were clearly worked out for all stakeholders, won over the attendees and potential investors. Blockchain as a technology component was also recognized and appreciated as a suitable technology. Marc had invited Linda to the presentation of the business idea and presented her as a health expert. The start-up ultimately was able to realize another round of financing and create a lot of attention in the community. The matter is clear to Marc and his team. They continue to live their mindset, to iterate the next set of functionalities in their MVP fast and with agility, and test the functions on the MVE. The goal is to generate more capital for the venture in a couple of months with an Initial Coin Offering (ICO).

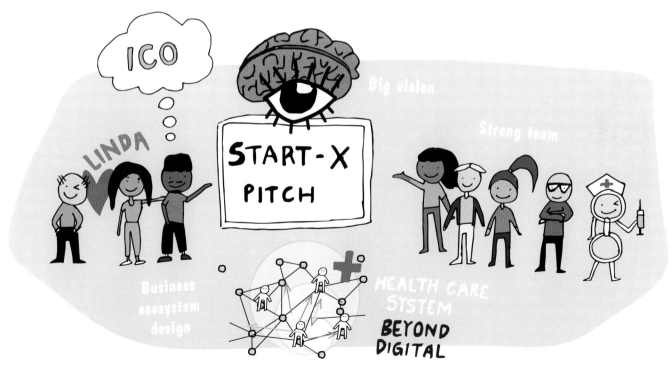

THANK YOU!

to Jana, Elena, Mario, Daniel, Isabelle …

Our thanks go to the many experts who have highlighted many facets, tools, and extensions of design thinking so that this *Design Thinking Playbook* could be produced for you.

WWW.DESIGN-THINKING-PLAYBOOK.COM

OVERALL KEY LEARNINGS:

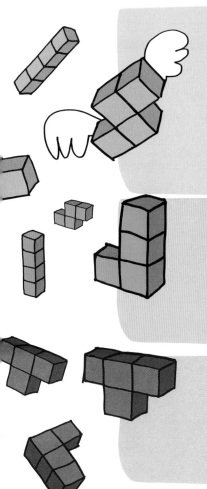

- Combine design thinking with systems thinking and the hybrid model—**even complex problems can be solved, agility heightened, and the range of solutions enlarged through the integration of various approaches!**
- Use the lean canvas to summarize the findings—**it is the link between the final prototype of design thinking and the lean start-up phase!**
- The design of business ecosystems becomes a key capability in networked structures—**think in value streams and win-win situations for all stakeholders to create a minimum viable ecosystem (MVE)!**
- New design criteria are essential in digitization. With the use of artificial intelligence and human–robot interaction, there is an exchange of information, knowledge, and emotions—**design this interaction consciously and accept that complex systems require more complex solutions!**

- Design not only the space but also the work environment. Make sure that the creative spaces are not overloaded —**less is more!**
- Put interdisciplinary teams together consisting of T-shaped and Pi-shaped members—**the transparency of the thinking preferences helps to build winning teams!**
- Create an organizational structure without silos and a mindset that matches the organization—this is the only way to disseminate design thinking transversally in the company.
- **Apply strategic foresight as the ability to plan and design the desired future—successful companies have a clear strategy and leaders who promote these visions!**

- Internalize the mindset and the design thinking process, work in short iterations, and develop an awareness of the groan zone—**it is critical in order to be successful in the end!**
- Build up empathy by understanding the actual needs and the background of potential users—**this is the only way to realize true innovations!**
- Create prototypes under time pressure and test them as early as possible in the real world. Integrate the various stakeholders in the testing—**the principle is: Love it, change it, or leave it!**

PRESENTATION OF THE EXPERTS

MICHAEL LEWRICK  PATRICK LINK

#Digitization #Combination of approaches #Design thinking　　　　#Agile development #Idea to scale #Design thinking

RÉSUMÉS

Michael has had different roles over the last few years. He was responsible for strategic growth, acted as Chief Innovation Officer, and laid the foundation for numerous growth initiatives in sectors that are in a transition.

He teaches Design Thinking as a visiting professor at various universities. With his help, a number of international companies have developed and commercialized radical innovations. He postulated a new mindset of converging approaches of design thinking in digitization.

Patrick has been Professor for Product Innovation at the chair for Industrial Engineering/Innovation at Lucerne University of Applied Science—Technology & Architecture since 2009. He studied Mechanical Engineering at ETH Zurich, then worked as a Project Engineer before receiving his doctorate in the field of innovation management at ETH Zurich. After eight years at Siemens, he now teaches product management and deals intensively with the advancement of agile methods in product management, design thinking, and lean start-up.

WHY ARE YOU A DESIGN THINKING EXPERT?

I came into contact with design thinking for the first time in Munich in 2005. At the time, it was a question of supporting start-ups in the development and definition of new products. In recent years, I attended to various company projects, seeing them through at Stanford University. In the context of my various functions in different industries, I was able to initialize a multitude of co-creation workshops with major customers, start-ups, and other actors in the ecosystem and thus advance various methods and tools.

When I first became acquainted with design thinking, I quickly realized the potential of this approach for interdisciplinary collaboration. Since then, we have used the approach in many training and advanced training modules. In particular, the combination of intuitive, circular approaches and analytical methods is very instructive. Together with colleagues from industry, we advanced design thinking and other agile methods and offer workshops and courses.

YOUR MOST IMPORTANT DESIGN THINKING TIP?

I know many experts who practice the prevailing design thinking approach with heart and soul and great commitment. For this very reason, we must constantly reflect upon and advance our design thinking mindset. New technologies and progressing digitization offer new opportunities for the development of ideas and the design of customer experiences. I have two tips: Use big data/analytics and systems thinking more extensively in the individual design phases, and integrate the new design criteria of digitization today in the development of innovations for tomorrow.

There is a danger—especially in inexact sciences—that experts proselytize and want to convince others of their approaches. The mindset and its adaptation to the respective context are more important than the process or method. Because all agile and lean approaches have basically the same mindset, you can learn a great deal from the other approaches and experts.

Try out the combination of design thinking with other approaches (hybrid model).

LARRY LEIFER

#Design thinking

NADIA LANGENSAND

#Visualization

RÉSUMÉS

Larry is a Professor of Mechanical Engineering Design and founding director of the Center for Design Research at Stanford (CDR) and of the Hasso Plattner Design Thinking Research Program at Stanford. He is one of the most influential personalities and pioneers of design thinking. He brought design thinking to the world, putting the focus on working in interdisciplinary teams.

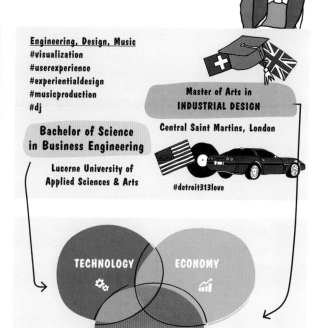

Engineering, Design, Music
#visualization
#userexperience
#experientialdesign
#musicproduction
#dj

Master of Arts in
INDUSTRIAL DESIGN

Bachelor of Science
in Business Engineering

Central Saint Martins, London

Lucerne University of
Applied Sciences & Arts

#detroit313love

WHY ARE YOU A DESIGN THINKING EXPERT?

I have been dealing with design thinking and the research in this area for decades. This includes global team dynamics, interaction design, and adaptive mechatronic systems.

In the ME310 program, I was able to observe the cultural differences in a variety of projects and use cases and infer significant conclusions from it for teaching and research at Stanford.

TECHNOLOGY

ECONOMY

DESIGN

YOUR MOST IMPORTANT DESIGN THINKING TIP?

Steve Jobs said it in a nutshell: "Think different!" The correct phrase, of course, would be, "Think differently!" With this, he had expressed the essence of design thinking: Don't necessarily do what is expected and what's understood as being the right thing. The design of human–robot interactions will gain in importance in the future. The emotional component must be emphasized more when defining the design criteria.

SHOW AND VISUALIZE

COURAGE TO DRAW AND QUICK 'N'
DIRTY PROTOTYPING

ARMIN LEDERGERBER

BEAT KNÜSEL

RÉSUMÉS

Since 2015, Armin Ledergerber has been a Service Designer with a focus on social media and cognitive computing in Customer Inter-action Experience at Swisscom, the leading telecommunications company in Switzerland. Prior to that, he was a project manager and research associate at the Institute for Marketing Management at Zurich University of Applied Sciences.

Beat is an interdisciplinary inventor, founder, and lecturer at various universities. His research areas are service excellence and business innovation. As a graduate electrical engineer with postgraduate studies in microelectronics, software design, and business administration, he has experienced the whole value chain from R&D to top management. With his consulting company ErfolgPlus, he accompanies companies in the digital transformati-on. In the startup TRIHOW he researches the use of smart, haptic aids in the design thinking environment.

WHY ARE YOU A DESIGN THINKING EXPERT?

I came into contact with design thinking when doing my master's degree, especially in the area of customer journey mapping and the development of personas. Since then, I have had many opportuni-ties to apply other methods of design thinking in various projects in order to design, test, and finally realize user-centric solutions iteratively.

I have always been fascinated by humans and technology. I have found deep chasm between people and technology in many companies. One talks a lot and understands little. Tangible proto-types in design thinking enable us to better understand, unleash creativity, and create magical team moments. A big concern for me is to bring back the haptics into the work processes. It is probably the most important ingredient in digital transformation and combines people, business, and culture.

YOUR MOST IMPORTANT DESIGN THINKING TIP?

Get out on the street and to the user or potential customer as early as possible. Ideally, you confront the user already with the first idea or concept outline so that inputs can be integrated in the next itera-tion. Put simply: Don't wait too long to venture into the real world.

Design thinking uses all the senses. Think again and again how you can optimally incorporate them. Be brave and quickly overcome the hurdle from thinking to doing. Use the space and be skillful with all the expressions that make your thoughts touchable. This way of working is the ideal track for the pathway from the ego to the we.

DANIEL OSTERWALDER

#2.5 Facilitation & visualization

DOMINIC HURNI

#1.5 Empathy with the users

RÉSUMÉS

As a facilitator, design thinker, and graphic recorder, he is out and about with his own company, Osterwalder & Stadler GmbH. In conjunction with companies, he developed various thinking labs in which participants also learn to persevere in the somewhat unfamiliar emergence of such laboratories. He is also known for his very impressive graphic recordings on meter-long paper strips.

Dominic teaches at Bern University of Applied Sciences, Innovation and Change Management, and is the founder of INNOLA GmbH, which addresses innovations for life in old age. His journey led him through a wide variety of occupations and locations: gardener in Ireland, care professional in Antigua in the Caribbean, instructor at the AIDS Federation in Bern, and process manager in the Swiss financial industry. In design thinking, Dominic saw an opportunity to deploy his T-shaped personality to create an impact.

WHY ARE YOU A DESIGN THINKING EXPERT?

As a facilitator and volunteer coach in professional sports, I'm mainly interested in the way the coaching and support for innovative processes and projects is done as well as the question of how peak performances can be achieved over and over. In addition, I teach design thinking in the context of facilitation change, so that the big picture of a transformation does not get lost in the shuffle.

I got to know design thinking during my master's degree studies in 2010. Since then, I have been practicing design thinking in a wide variety of projects and a wide range of areas, including the health care sector, insurance, and education. The challenge in the context of design thinking switched from a focus on methods to conveying the mindset in a zero-error culture, especially in large companies.

YOUR MOST IMPORTANT DESIGN THINKING TIP?

Dialog begins with listening. It is—as William Isaac said so wonderfully aptly—the art of thinking in common and—I'd like to add—acting in common. When you manage design thinking projects, you should be present but low key; be there, follow things with every fiber of your mind and body, and pay the utmost attention to how you can promote and encourage this way of thinking in common.

Human beings have two ears and one mouth. Thus the ratio is set: Listen twice as much as you speak. One big mistake: "I know what my customer wants. I don't have to ask him." Don't be afraid of people's eyes rolling and the feeling you're the pollo (Spanish for chicken). Only those who ask questions will get an answer and the opportunity to learn. Another mistake: "My customer can say to me what he wants." Customers usually express what they would like or what gets them angry. Only rarely do they suggest an innovative solution. Empathy is the key.

327

ELENA BONANOMI

#1.6 Correct focus & 360 degrees

EMMANUEL SAUVONNET

#3.2 Business models & innovation

RÉSUMÉS

As a partner of Innoveto, Elena supports her customers in their innovation projects. Time and again, she comes up with new ideas to make the journey instructive and inspiring. Design thinking and agile innovation form the core aspects and are enriched with her own inspirations. Previously, Elena produced fresh ideas with BrainStore and learned to switch on the ideas machine for the challenges of customers.

Emmanuel is the founder and managing partner of neueBerating GmbH, a leading consultancy firm for the development and expansion of new business fields and business models. Ever since graduation, he has been passionately involved in the development and implementation of new business models.

WHY ARE YOU A DESIGN THINKING EXPERT?

Since 2013, design thinking has been my professional life. I support clients in their innovation processes using approaches from various disciplines. In addition, I design new projects in which I translate theoretical knowledge into practical skills. I discovered the design thinking discipline when doing my master's and was quickly totally taken with it. I have deepened my knowledge analyzing several scientific papers.

I experienced design thinking initially in 2011. I was thrilled that so many of the individual components I was already familiar with were embedded in an overarching concept and thus developed an entirely new momentum. Basically, design thinking combines many existing elements and links them successfully. This is why it is more than just a pure innovation tool.

YOUR MOST IMPORTANT DESIGN THINKING TIP?

I'm often asked whether it's not a bit risky to deal with ideas so openly and transparently. An idea is not a product yet, nor a working business model. In order to get there, a process must be carried out in which phases of development alternate with phases of testing and in which both the users and the stakeholders are actively involved. Feedback and active contributions are important. This is the only way for the idea to become a product that meets real needs and delivers a clear value for the user.

Suspending the hierarchy in the design thinking process is vital for a good result. It's not always easy to achieve it, though. The difficulty is not only to get executives to accept it but also that employees need to learn to develop and communicate their own ideas. They must stop looking out for signals that come from their supervisors. Both sides must be balanced during the workshops.

ISABELLE HAUSER

JAN-ERIK BAARS

RÉSUMÉS

Isabelle is a lecturer at HSLU, Engineering & Architecture, and teaches at the interface between technology, economic feasibility, and design. As an industrial designer with her own agency, she advises and supports customers and SMEs in technical areas with industrial design solutions as well as with in-house workshops for design thinking and product developments.

Jan-Erik is currently teaching the Design Management course at Lucerne University of Applied Science. He is also active as a consultant and supports companies with the effective and comprehensive use of design. He writes and speaks about his experience in the world of design and tries to promote human-centric design and the acceptance of design as a key component of corporate excellence.

WHY ARE YOU A DESIGN THINKING EXPERT?

As an industrial designer, the leap to design thinking is not so big. Various courses for further training, conferences, and stays in Stanford as well as my work at the d.school there helped me to build up a profound knowledge in the subject matter. Since then, I have been responsible for international, interdisciplinary study teams with partners from industry. In numerous workshops that I hold, I let entrepreneurs experience and get an understanding of design thinking.

I learned design thinking from scratch after I'd worked for a long time as a designer at Philips. One of my tasks was to gear the development of products to the needs of people. One thing I learned there is that a company is received well by the customer only if the collaboration of all those involved in the company is top-notch, along the lines of: "What goes around, comes around!"

YOUR MOST IMPORTANT DESIGN THINKING TIP?

The "bias toward action" mindset is not only a recommendation but a must. If you don't experience design thinking yourself, you won't be able to estimate its benefits. Not infrequently, I have had the pleasure of getting positive feedback even from the most stoic of skeptics after they've taken part in one of my workshops. The reason it works so well is that design thinking has something to offer everybody.

As a consultant, I experience the introduction of design thinking firsthand. It is often taken up by committed decision makers and introduced to the company, not infrequently by the boss himself. Unfortunately, further implementation is often thwarted by the lack of expertise in the greater organization. Hence my tip is always to clarify the current situation of the organization when launching design thinking. That can be done by examining the degree of customer centricity.

JANA LÉV

#2.2 Interdisciplinary teams

MARIO GURSCHLER

#1.10 Efficient testing & digital tools

RÉSUMÉS

Jana works as an Innovation Manager and Senior Management Consultant in "Die Mobiliar," a property insurer. Together with an interdisciplinary team, she develops new products and services there. In addition, she lectures at different universities on design thinking and innovation and speaks at conferences on how the transformation into a customer-oriented company can be successfully done.

Mario has been working as a Product Marketing and Business Development Manager since 2000. He has actively designed the portfolios of start-ups and multinational corporations in the tech industry to get them into peak condition to meet the requirements of e-commerce and digital transformation. As an industrial engineer, he is well versed in the mediation between business and technology, which has proven so important in design thinking.

WHY ARE YOU A DESIGN THINKING EXPERT?

I learned design thinking as a structured method at the d.school in Potsdam. Even before that, I provided customers with cardboard, scissors, and glue to develop their living environment in common with them. I was able to deepen my expertise in my functions as change agent in a Web agency and as an advisor of various corporations, which wanted to become more customer-oriented and more innovative.

As a product manager, I experienced myself how solutions focusing on the technology have become less and less attractive and that investment decisions are no longer made by those responsible for technology. For me, design thinking as a method solved this paradox because it puts the people and the problem you want to solve for them at center stage. Design thinking belongs in everybody's skill set toolbox.

YOUR MOST IMPORTANT DESIGN THINKING TIP?

Design thinking wins over employees when it can be experienced in actual practice. The added value of empathy, interdisciplinary collaboration, and fast, iterative prototyping along the lines of "fail early and often" will swiftly become clear. Anybody interested in applying design thinking to their own company/project should try to experience it as quickly and simply as possible (e.g., in courses/workshops).

In many ways, design thinking breaks with putatively proven approaches. This is why it often meets with resistance. To scribble ideas with colored pencils and paper, instead of presenting PowerPoint slides, is quickly seen as unprofessional or even esoteric. Don't give up! The result will convince even the doubters.

MARKUS BLATT

#3.2 Lean models & business models

MIKE JOHNSON

3.1 Systems thinking

RÉSUMÉS

Markus is the founder and managing partner of neueBerating GmbH and has been a business consultant for 15 years. As a student, he founded a start-up and gained invaluable experience in e-commerce. Online topics and innovative business models are still part and parcel of his work as a consultant: He has designed the digital transformation of many companies in the imaging and media industry. A graduate in business administration, he nonetheless feels at home among designers.

Mike has been living in Switzerland since 2011. He works as a Systems Engineering Manager in the health care industry. He also has experience in the aerospace and defense industry. He is a co-founder of the Swiss Society of Systems Engineering (SSSE) and of the Council on Systems Engineering (INCOSE). His great passion is product development, and his track record in the implementation of complex projects is impressive indeed.

WHY ARE YOU A DESIGN THINKING EXPERT?

In my projects, I make sure I follow design thinking. Because only when technology, economic feasibility, and the alignment come together at the target group can real innovations emerge. I implement this in a form of project work, which is composed of workshops and open-innovation elements, and test the planned project results with the target group early on.

I'm not an expert in design thinking! My strength lies in the field of systems engineering. Some time ago, though, I realized there are significant overlaps between design thinking and systems engineering. Over the course of my career, I realized many complex projects. This experience helps me to educate and train others and to advance the approach through professional organizations such as the SSSE.

YOUR MOST IMPORTANT DESIGN THINKING TIP?

Fail early! An excellent way is to build a 80% prototype with the least effort and then test it with the target group. I'm a perfectionist, so this part is always a little difficult for me. But it helps to keep up the speed and prevents hitting a dead end early on.

In complex product development projects, the importance of the engineering team should not be underestimated. People are the most complex systems in the universe that we know. For this reason, systems made up of people (e.g., teams) are even more complex. To guide such teams in a targeted way, it's necessary to have a team leader or facilitator with excellent communication and social skills.

NATALIE BREITSCHMID

#1.9 Prototyping, #2.1 Spaces & environment

NATALIE JÄGGI

RÉSUMÉS

Natalie likes to tap unknown fields of application with design thinking, and for several years has been responsible for teams in human-centric service and product design at Swisscom. She is certainly not above any feedback debacle in terms of a half-baked prototype. She induces enthusiasm in others, with whom she loves to experiment with wild and free thoughts. Since the turn of the millennium, she has applied and taught the FlowTeam® method.

Natalie is the Head of Service Experience at Swisscom. She worked as a research assistant at the Institute for Business Informatics at Bern University and did her PhD in the area of online marketing.

WHY ARE YOU A DESIGN THINKING EXPERT?

Since 2010, I have been deeply involved in design thinking. I began coaching and experimenting in my own projects with a curious attitude, a balanced creative/analytical approach, and inspiring tools. I achieved breakthroughs in service design that still today inspire me and many other people. What mainly interests me are new, as yet unknown areas of application.

As early as when writing my doctorate, I dealt with customer retention and customer behavior. When I started at Swisscom, I encountered design thinking for the first time in a human-centered design workshop. Design thinking is a priority at Swisscom, and as an employee you get a great deal of support in the use of the methods. Everybody here should have the opportunity of using design thinking in their day-to-day work.

YOUR MOST IMPORTANT DESIGN THINKING TIP?

Design thinking fascinates me because intuitive and analytical skills can be combined with each other. Sometimes, this results in participants being completely at a loss during the process. This is why I find the aspect of letting go quickly quite intriguing—when testing a prototype, for instance. If you can overcome the feeling of failure as rapidly as possible and transform it positively, you will achieve a satisfying goal far more quickly.

No matter the idea you have and no matter how excited you and your colleagues are over it, the idea is valuable only once it has been tested with the customer. Begin involving the customer in the early phases of your ideas. His or her feedback ought to provide you with guidance. Usually, the first prototype is enough to get exciting statements.

SOPHIE BÜRGIN

#1.4 Discover needs & needfinding

STIJN OSSEVOORT

#2.4 Storytelling

RÉSUMÉS

Sophie works as a user insight expert for the Bern-based innovation consultancy INNOArchitects and also as a self-employed consultant. Sophie has international project experience in design thinking, service design, and lean projects, which she attended to and saw through in many industries as well as in the social environment. She is a lecturer in design thinking. She attaches great importance to making the mindset tangible.

Stijn is a lecturer at the Lucerne University of Applied Science. He does research in the areas of user-centric design, sustainability, and product interaction. He participates in the Competence Center Visual Narratives, where he is considered an expert for interfaces between users and products. As a passionate designer, he is busy confronting the world with unique and functional prototypes.

WHY ARE YOU A DESIGN THINKING EXPERT?

During my time at Deutsche Bank, I was trained in the Stanford design thinking method and the foresight thinking method. The perspective as an outside-in design thinker taught me to attend to and accompany the team dynamics of multidisciplinary innovation teams. As a design ethnographer, I specialized in handling people and in methods for needfinding and user insights.

Design has a social character. True innovations do not come into being behind a computer screen—they result from the social interaction outside our comfort zone. My expertise in design thinking is to create simple prototypes or visual ideas, with which the user can interact, and infer from them the right conclusions.

YOUR MOST IMPORTANT DESIGN THINKING TIP?

Only too often are decisions made in day-to-day work based on assumptions. Design thinking teaches us the conscious dealing with assumptions. It often seems hard to venture out into the world of the user. Be courageous—people have exciting stories to tell that might enlarge your ideas by worlds. The only way to get better in needfinding is to do it yourself—the best place to learn it is outside, in the everyday life of people.

Get feedback in your project as often as possible and engage in a dialog with the users and potential customers. And feedback is more than just a collection of opinions. It is a tool with which you learn to understand your own ideas and goals better. I always give my students the tip of presenting their prototypes to close family members. In my opinion, this is the core of design thinking: involve and collaborate!

TAMARA CARLETON

#2.7 Strategic foresight

THOMAS EPPLER

Design thinking methods

RÉSUMÉS

Tamara is the CEO and founder of the Innovation Leadership Board. She develops tools and processes that promote radical innovation. She was a fellow at the Bay Area Science & Innovation Consortium and previously worked as a consultant at Deloitte Consulting. She specializes in customer experience, marketing strategy, and innovation. She has a PhD and is often invited to lectures to present her work and research.

Thomas works as a Senior Innovation Manager at SIX Payment Services and teaches innovation methods. He studied philosophy and cultural anthropology and has a Master of Advanced Studies in Innovation Engineering. Today, his focus is on creating meaningful and strong added value for customers. He develops and programs digital solutions in the areas of medicine, finance, industrial, and HR. Recently, he won an international hackathon.

WHY ARE YOU A DESIGN THINKING EXPERT?

I've always been keenly interested in the question of how ideas are developed in order to fulfill human needs. I started my career at the Center for Design Research at Stanford University. I was part of a global research community there. Over the last two years, I have enlarged the SUGAR network and turned it into a global innovation network for academic institutions that cooperate with companies in order to solve real-world problems.

I have a deep understanding of end users. Since the 1990s, my way of working has been characterized by creative thinking and the fast development of prototypes. I didn't imagine that there would be a name for it later. Several times, I led entire teams toward agile working, customer focus, and design thinking. I teach blue ocean strategy, lead-user method, and design thinking at various technical colleges.

YOUR MOST IMPORTANT DESIGN THINKING TIP?

Interesting solutions begin with interesting problems. Keep reformulating the problem. The foresight tools we talked about help understanding the problem better, capturing the problem space, finding unexpected solutions, and supporting the thinking on the team.

Take a lot of time for dealing with end users. The really intriguing and innovation-relevant insights are often only gained once trust has been built after quite some time and the other person is more open. Don't reduce a personal, in-depth exchange to a 10-minute interview. Don't turn a conversation into an interrogation. And never ever replace personal contact with a phone call.

WILLIAM COCKAYNE

#2.7 Strategic foresight

RÉSUMÉS

Bill is the CEO and founder of Lead|X, a learning platform that conveys business knowledge in short lectures via social media. As an entrepreneur and interim manager, he has managed various teams and successfully completed 20 projects. At Stanford University, he teaches Foresight & Technological Innovation. He has a PhD, has registered various patents, and wrote the "Playbook for Strategic Foresight and Innovation."

WHY ARE YOU A DESIGN THINKING EXPERT?

If you want to look actively into the future in spite of all the uncertainties, the golden rule is to look for the changes actively. Over the last 25 years, I have invented, designed, built, and delivered various products. From this experience, I realized that design is elementary for the entire innovation process and that a user-centric approach is vital for success.

YOUR MOST IMPORTANT DESIGN THINKING TIP?

Each change brings opportunities. Those who recognize the change early on can implement results, solutions, and ideas or at least create the necessary acceptance for it. Seeing the world as constantly changing allows us to get a better grasp of the future. If you aim at the near future, you can shape it and thus attain a competitive advantage.

SOURCES

Sources

Blank, S. G. (2013): Why the Lean Start-Up Changes Everything. Harvard Business Review, 91(5), pp. 63–72.

Blank, S. G., & Dorf, B. (2012): The Start-Up Owner's Manual: The Step-by-Step Guide for Building a Great Company. Pescadero: K&S Ranch.

Brown T. (2016): Change by Design. Wiley Verlag.

Buchanan, R. (1992): Wicked Problems in Design Thinking. Design Issues, 8(2), pp. 5–21.

Carleton, T., & Cockayne, W. (2013): Playbook for Strategic Foresight and Innovation. Download at: http://www.innovation.io

Christensen, C., et al. (2011): The Innovator's Dilemma. Vahlen Verlag.

Cowan, A. (2015): Making Your Product a Habit: The Hook Framework. Accessed Nov. 2, 2016, http://www.alexandercowan.com/the-hook-framework/

Davenport, T. (2014): Big Data @ Work: Chancen erkennen, Risiken verstehen. Vahlen Verlag.

Davenport, T. H., & Patil, D. J. (2012): Data Scientist: The Sexiest Job of the 21st Century. Harvard Business Review, Oct. 2012 https://hbr.org/2012/10/data-scientist-the-sexiest-job-of-the-21st-century/

Gerstbach, I. (2016): Design Thinking in Unternehmen. Gabal Verlag.

Herrmann, N. (1996): The Whole Brain Business Book: Harnessing the Power of the Whole Brain Organization and the Whole Brain Individual. McGraw-Hill Professional.

Hsinchun, C., Chiang, R. H. L., & Storey, V. C. (2012): Business Intelligence and Analytics: From Big Data to Big Impact. MIS Quarterly, 36(4), pp. 1165–1188.

Kim, W., & Mauborgne, R. (2005): Blue Ocean Strategy. Hanser Verlag.

Leifer, L. (2012a): Interview with Larry Leifer (Stanford) at Swisscom, Design Thinking Final Summer Presentation, Zurich.

Leifer, L. (2012b): Rede nicht, zeig's mir. Organisations Entwicklung, 2, pp. 8–13.

Lewrick, M., & Link, P. (2015): Hybride Management Modelle: Konvergenz von Design Thinking und Big Data. IM+io Fachzeitschrift für Innovation, Organisation und Management (4) pp. 68–71.

Lewrick, M., Skribanowitz, P., & Huber, F. (2012): Nutzen von Design Thinking Programmen, 16. Interdisziplinäre Jahreskonferenz zur Gründungsforschung (G-Forum), University of Potsdam.

Lewrick, M. (2014): Design Thinking–Ausbildung an Universitäten. In: Sauvonnet and Blatt (eds.), Wo ist das Problem? pp. 87–101. Neue Beratung.

Link, P., & Lewrick, M. (2014): Agile Methods in a New Area of Innovation Management, Science-to-Business Marketing Conference, June 3-4,

2014, Zurich, Switzerland

Maurya, A. (2010): Running Lean: Iterate from Plan A to a Plan That Works. The Lean Series (2nd ed.). O'Reilly.

Moore, J.F. (1993): Predators and Prey: A New Ecology of Competition. Harvard Business Review, 71, pp. 75–86.

Moore, J.F. (1996): The Death of Competition: Leadership & Strategy in the Age of Business Ecosystems. HarperBusiness.

Ngamvirojcharoen, J. (2015). Data Science + Design Thinking. Thinking Beyond Data – When Design Thinking Meets Data Science. Accessed Dec. 12, 2015, http://ilovedatabangkokmeetup.pitchxo.com/decks/when-design-thinking-meets-data-science

Norman, D. (2016): The Design of Everyday Things: Psychologie und Design der alltäglichen Dingen. Vahlen Verlag.

Oesterreich, B. (2016): Das kollegial geführte Unternehmen: Ideen und Praktiken für die agile Organisation von morgen. Vahlen Verlag.

Osterwalder, A., et al. (2015): Value Proposition Design. Campus Verlag.

Osterwalder, A., & Pigneur, Y. (2011): Business Model Generation. Campus Verlag.

Porter, M. E., and Heppelmann, J. E. (2014): How Smart, Connected Products Are Transforming Competition. Harvard Business Review 92(11), pp. 11–64.

Ries, E. (2014): Lean Startup: Schnell, risikolos und erfolgreich Unternehmen gründen. Redline Verlag.

Sauvonnet, E., & Blatt, M. (2014): Wo ist das Problem? Mit Design Thinking Innovationen entwickeln und umsetzen, 2nd ed., 2017. Vahlen.

Savoia, A. (2011): Pretotype it. Accessed Jan. 2018, http://www.pretotyping.org

Schneider, J., & Stickdorn, M. (2011): This Is Service Design Thinking. Basics – Tools – Cases. BIS Publishers.

Siemens (2016): Pictures of the Future, Accessed Nov. 1, 2016, http://www.siemens.com/innovation/de/home/pictures-of-the-future.html

Szymusiak, T. (2015): Prosumer – Prosumption – Prosumerism. OmniScriptum GmbH & Co. KG, pp. 38–41.

Uebernickel, F., & Brenner, W. (2015): Design Thinking Handbuch. Frankfurter Alllgemeine Buch.

Ulwick, A. (2005): What Customers Want: Using Outcome-Driven Innovation to Create Breakthrough Products and Services. McGraw-Hill Higher Education.

Vandermerwe, S., & Rada, J. (1988): Servitization of Business: Adding Value by Visionaries, Game Changers, and Challengers. Wiley.

von Hippel, E. (1986): Lead Users. A Source of Novel Product Concepts. Management Science, 32, pp. 791–805.

INDEX

INDEX

N

O

P

Notes

Notes

Notes

Notes

Notes